Sing Yiu Cheung

Traffic Surveillance by Wireless Sensor Networks

Sing Yiu Cheung

Traffic Surveillance by Wireless Sensor Networks

Vehicle Detection, Classification and Re-Identification by Wireless Sensor Networks

VDM Verlag Dr. Müller

Imprint

Bibliographic information by the German National Library: The German National Library lists this publication at the German National Bibliography; detailed bibliographic information is available on the Internet at http://dnb.d-nb.de.

Cover image: www.purestockx.com

Publisher:
VDM Verlag Dr. Müller Aktiengesellschaft & Co. KG , Dudweiler Landstr. 125 a, 66123 Saarbrücken, Germany,
Phone +49 681 9100-698, Fax +49 681 9100-988,
Email: info@vdm-verlag.de

Zugl.: Berkeley, The University of California, 2007

Produced in USA and UK by:
Lightning Source Inc., La Vergne, Tennessee, USA
Lightning Source UK Ltd., Milton Keynes, UK
BookSurge LLC, 5341 Dorchester Road, Suite 16, North Charleston, SC 29418, USA

ISBN: 978-3-639-01559-1

Abstract

Traffic surveillance technologies are the foundations of Intelligent Transportation System (ITS), which provide part of, or even all data used by other ITS systems. Because of the highly intrusive property of inductive loop detectors, the quest for researching a reliable and cost-effective alternative system has been underway for some time. Wireless sensor networks based traffic surveillance systems are proposed to replace inductive loop systems, and extend the coverage of ITS applications over all freeways and local intersections.

In this dissertation, the development and evaluation of a traffic surveillance system by wireless sensor networks are presented. The hardware and software specifications of the sensor node prototypes, as well as its communication protocols and lifetime analysis are provided. Signal analyses of acoustic and magnetic sensors on vehicles are studied. An efficient and robust real-time vehicle detection algorithm for these sensors, called Adaptive Threshold Detection Algorithm (ATDA) is developed. Detection rate higher than 97% is achieved in real-time experiments.

1

Vehicle classification and reidentification schemes targeting at platforms with very limited computation resources are developed. Novel transformation processes called Average-Bar and Hill-Pattern transformation are used to compass the vehicle signatures for further processing. Vehicle classification experiment that apply k-Nearest Neighbor (k-NN) and Support Vector Machine (SVM) classifiers are presented. For vehicle reidentification, a Max-Of-Max (MOM) algorithm is proposed to use with the Average-Bar compassed signatures. A left-turning reidentification experiment is presented.

The feasibility for large scale deployment of these wireless sensor networks has the potential to revolutionize the whole traffic surveillance and control industry. Its impacts on a number of Intelligent Transportation Systems (ITS) are studied.

Table of Contents:

List of Figures:

List of Tables:

xiv

Acknowledgements

Throughout the course of this research project, I have been working and interacting with a wide range of people. Having the opportunity to work with them is one of the most valuable experiences of my research. While I cannot thank each of them individually in writing, I would like to acknowledge some of them who have been particularly helpful and inspiring.

Professor Pravin Varaiya is the founder of this research project, as well as a famous pioneer in the Intelligent Transportation System (ITS) industry. This project was born because of his intellectual insight into the next generation of ITS technologies. His never-ending interest in work and inspiring directions have sustained the project. I warmly appreciate his enthusiastic support throughout these years.

Professor Roberto Horowitz has graciously acted as a co-advisor on my research. I am very thankful for his advice and patience on me since the very beginning of my study. My colleagues Chin-Woo Tan, Jiagen Ding, Sinem Coleri and Baris Dundar have been a tremendous support throughout this work. I am very grateful to be in the same research team with them.

My most heartfelt gratitude goes to my parents, sister and girl friend for their constant love and confidence.

Ch. 1 Introduction

1.1 Purpose and Importance of the Research

Increasing traffic congestion is a critical problem in many countries. Between 1980 and

1998, vehicle miles traveled increased 72% in the U.S. while the number of lane miles

increased only 1% [7.20]. The 2005 Urban Mobility Report [1.1] shows that the total cost

of congestion for the 85 U.S. urban areas is estimated to be 65 billion dollars per year,

which come from 3.5 billion hours of delay and 5.7 billion gallons of excess fuel

consumed. In order to prevent congestion problem from getting worse, the U.S.

government initiated the Federal Intelligent Transportation System (ITS) program in 1991

for the development and deployment of advanced technologies for maximizing the traffic

capacity and minimizing the delay. However, the current pace of improvement in

transportation system is not sufficient to keep pace with even a slow growth in the traffic

demands in most major urban areas.

Advanced management subsystems in ITS, such as traveler information systems, freeway

and arterial management systems, emergency management, and parking management,

increasingly rely on monitoring of real-time traffic network conditions [section 7.1].

Traditional transportation management divisions such as transportation planning and

pavement maintenance also need the associated traffic data. For instance, the Traffic

Management Center (TMC) can optimize the cycle time of traffic lights based on queue

1

lengths [section 7.1]. Travelers can use this information to plan their activities and routes. There is a great need for advanced surveillance capabilities to complement the rapid deployment of ITS strategies. Since the quality of traffic data influences the proper functions of the ITS systems, the data collected must be plentiful, diverse, and accurate, which presents a serious challenge to the traffic surveillance industry [section 2.1.1].

Most conventional traffic surveillance systems use intrusive sensors, which include inductive loop detectors [1.2] [1.3], micro-loop probes, pneumatic road tubes, piezoelectric cables and other weigh-in-motion sensors [section 2.1.1]. They are chosen because of their high accuracy for vehicle detection (> 97%). For maximizing the benefits from all these ITS technologies, a large scale deployment of traffic controls on all major freeways and local streets would be an eager step to be taken [1.1]. Therefore, real-time traffic information at all these sites is required. However, serious disruption of traffic is induced by the installation and maintenance of surveillance system, which leads to a relative high system cost at the level of ten thousand dollars per intersection [section 2.1.2]. Therefore, these systems are not cost-effective enough for large scale deployment.

In this research project, wireless sensor networks were developed and implemented as a traffic surveillance system with detection accuracy as good as that of inductive loop detectors [section 4.4]. They offer a very attractive alternative to inductive loops for traffic surveillance. The sensor networks have a much higher configuration flexibility, which makes the system scalable and deployable everywhere in the traffic network. The availability of these data opens up new opportunities for intelligent traffic operations and

2

control [Ch. 7]. With a much lower system life-cycle cost than inductive loop, video and radar detector systems [section 2.1.2], sensor networks are cost-effective for large scale deployment. It has the potential to revolutionize the whole traffic surveillance and control industry [Ch. 7].

1.2 Summary of the Dissertation

Background and Motivations

Traffic surveillance technologies are the foundations of Intelligent Transportation System (ITS), which provide part or all of the data used by other ITS systems. The surveillance technologies are updated constantly to provide historical and real-time data of traffic count, speed, classification and re-identification. In order to provide a better understanding on the background of the traffic surveillance industry that this research project is focusing on, a review of traffic surveillance technologies and motivation for using wireless sensor networks are presented in chapter 2.

Surveillance technologies can be classified into three categories: intrusive, non-intrusive and off-roadway technologies. Intrusive traffic sensors are installed within or across the pavement. Non-intrusive sensors can be installed above or on the sides of roads with minimum disruption to traffic flow. Off-roadway technologies refer to those that do not need any specific equipment to be installed at the test site. In section 2.1.1, several state-of-the-art technologies in each of these categories are studied. Comparisons of their data type availability; system performance; and cost are presented in section 2.1.2.

Because of the highly intrusive property of inductive loop detectors, the quest for researching a reliable and cost-effective alternative system, which can provide traffic data at the same accuracy level as inductive loop systems, while minimizing the disruption during installation and maintenance, has been underway for some time. The motivation for developing wireless sensor networks based surveillance system is to provide a direct replacement for the inductive loop systems, and extend the coverage of ITS applications over all the freeways and local intersections. Such a large scale deployment has the potential to revolutionize the whole traffic surveillance and control industry. The characteristics of sensor networks that make it scalable and deployable everywhere are presented in section 2.2.

Wireless Sensor Networks

A Wireless Sensor Network (WSN) [3.1] is a network of small sensor nodes (SN) communicating among themselves using wireless communication, to sense the physical world. It combines distributed sensing, computation and wireless communication technologies. Conditions such as temperature, sound, vibration, pressure, motion or pollutants could be monitored on a large scale using a spatially distributed WSN (from tens to thousands of sensor nodes). Thanks to the revolution in sensor, processor, communication and power technologies, sensor nodes can now be integrated into a small millimeter-cubic size at a low cost [3.3]. Such technology advances push WSN into a new era as it is now flexible and cost-effective enough to be deployed on a large scale. In chapter 3, the architecture and components of a WSN, how it can be used in the traffic surveillance industry, the corresponding hardware and software specifications of the

4

prototypes that were developed, as well as the communication protocols and lifetime analysis are provided.

Vehicle Detection

Traffic surveillance starts with detecting the presence of a vehicle. Based on such detection, statistics like vehicle counts, traffic flow, speed and occupancy are calculated. This very first stage in the surveillance system determines the final performance of all the dependent applications. In chapter 4, signal analyses of two types of sensor, acoustic and magnetic, which can potentially be used in sensor nodes are presented in section 4.1 and 4.2 respectively. An efficient and robust real-time detection algorithm for these sensor signals, called Adaptive Threshold Detection Algorithm [4.1] is studied in section 4.3. Lastly, the experimental results and analysis of this vehicle detection system are presented in section 4.4.

Vehicle Classification

Vehicle classification refers to the process and methodology to classify a vehicle signature in a specific format into a pre-defined vehicle class (e.g. passenger vehicle or truck). It is an important source of information for transportation design and management that can be used in many contexts. In pavement design and management, pavement life is estimated according to the distribution of vehicle types running over it and the re-surfacing can be scheduled in advance. In traffic safety research and implementation, the distribution of trucks in traffic is a critical design factor, because of its significantly low speed and large size. In traffic control, signal priority can be given to vehicle classified as bus or emergency

5

vehicle. It also provides valuable data input to the prediction of highway capacity, assessment of the effectiveness of traffic legislation, automatic toll collection, weight enforcement strategies and environmental impact studies.

Similar to vehicle detection, several technologies were developed for classification. Vision-based, inductive loop, microwave, piezo-electric and acoustic-based classification technologies are the state-of-the-art techniques [section 2.1]. Vision-based classification can achieve a correct classification rate higher than 90% [5.1]. The major limitation of vision-based classification is that its performance is greatly affected by the environmental and lighting conditions. Classification stations with highly calibrated inductive loops are also being used [5.7] [5.8]. However, the infrastructure and maintenance cost of such a vehicle classification station is expensive. This makes the system economically justifiable only when being deployed at a particular site of interest (e.g. toll plaza). On the other hand, vehicle classification by wireless sensor networks provides a much more flexible deployment configuration, making the system portable and again scalable for large scale deployment.

In chapter 5, state-of-the-art classification technologies are first reviewed in section 5.1. The characteristics of magnetic vehicle signatures are studied in section 5.2. The data processing and classification schemes for a platform with very limited computation resources are presented in section 5.3. And the experimental results and analysis are presented in section 5.4.

Vehicle Reidentification

Vehicle reidentification is the process of matching the detections of a vehicle at different locations. In other words, it is the tracking of vehicles from point to point along a traffic surveillance network. A single detector station gives point traffic parameters such as speed and occupancy at that specific location. These measurements are assumed to be representative for the traffic extended from this station. However, this assumption is usually not valid when the traffic becomes congested. The limitation of point data raised the interest in vehicle reidentification, which provides section measurements including travel time, travel time variability, section density and origin/destination demand [6.1].

The section travel time and density are essential parameters for effective traffic planning, control and managements. The traffic parameters obtained from vehicle reidentification are used in evaluating the performance of dynamic traffic systems and act as input to advanced transportation management and information systems (ATMIS) to ease traffic congestion and its associated impacts [6.2]. In chapter 6, state-of-the-art technologies for section measurements and reidentification are reviewed in section 6.1. The proposed data processing and Max-Of-Max (MOM) reidentification scheme are provided in section 6.2. The corresponding experimental analysis and results are presented in section 6.3.

Evolution of Intelligent Transportation System

In the 2005 Urban Mobility report [1.1], the benefits from the implementation of four ITS technologies are studied: traffic signal coordination, arterial street access management, freeway entrance ramp metering and freeway incident management. The benefits are estimated to be 336 million hours of delay reduction and $5.6 billion in congestion savings for the 85 urban areas in 2003. If these technologies were deployed on all the major roads, an estimated 613 million hours of delay and more than $10.2 billion would be saved.

However, the large scale deployment of ITS technologies is discouraged by the high life-cycle cost [section 2.1.3] and huge traffic delay caused by the installation of inductive loop detectors. The proposed wireless sensor networks provide the same functionalities as inductive loop systems, but at a much lower cost and delay. Its wireless capability and configuration flexibility [section 2.2] give extra advantages to the large scale deployment of many ITS systems. Examples of traffic signal control, on-ramp metering and parking management are presented in section 7.1.

A multi-function wireless surveillance system can be developed by adding other sensing modalities to the traffic surveillance systems. An important modality for sensing road conditions is presented in section 7.2.1. The wireless communication capability of the surveillance system also allows it to talk to other ITS systems. Since the sensor nodes are located on the pavement, the networks can be a very useful tool in the Vehicle-Infrastructure Integration (VII) framework. The wireless communication can be used to exchange information between different systems and extend the vehicle-infrastructure communication range. Its applications to VII are presented in section 7.2.2.

Contributions and Future Developments

8

The proposed wireless sensor networks have the potential to revolutionize the whole traffic surveillance and control industry into one that is scalable and deployable everywhere in the traffic network. A summary of contributions of this research project is presented in section 8.1. And several potential future developments of this system are presented in section 8.2, including energy harvesting, installation-in-motion, multi-function networks and real-time implementation.

Ch. 2 Background and Motivation

For a better understanding of the background of this project, several common traffic surveillance technologies are studied in section 2.1. The motivations for using wireless sensor networks for traffic surveillance are provided in section 2.2.

2.1 Review of Traffic Surveillance Technologies

Traffic surveillance technologies form the foundations of Intelligent Transportation Systems (ITS), which provide part or all of the data used by other ITS systems. The surveillance technologies are updated constantly to provide historical and real-time data of traffic count, speed, classification and re-identification. No single surveillance system is best for all applications. Each has its own limitations, specializations, and capabilities.

In order to provide a better understanding on the background of the traffic surveillance industry that this research project is focusing on, the mechanisms and characteristics of several common traffic surveillance technologies are reviewed in section 2.1.1. Comparisons of their data type availability, system performance, and cost are presented in section 2.1.2.

2.1.1 Mechanisms of Different Surveillance Technologies

Surveillance technologies can be classified as intrusive, non-intrusive and off-roadway technologies. Intrusive traffic sensors are installed within or across the pavement. Non-

10

intrusive sensors can be installed above or on the side of roads with minimum disruption to traffic flow. Off-roadway technologies do not need any specific equipment to be installed at the test site. In this section, several common technologies in each of these categories are studied.

2.1.1.1 Intrusive Technologies

Intrusive technologies refer to those that require installation to be done directly onto the pavements, in saw-cut, holes or tunneling under the surfaces. Drawbacks include the disruption of traffic for installation and repair, failures induced by poor road conditions, and system reinstallation caused by road repairs or resurfaces. Examples include inductive loop, pneumatic road tube, piezoelectric cable, and weigh-in-motion system.

Inductive Loop

Inductive loop detector is the most common vehicle detector used in the traffic surveillance industry. Its basic setup is shown in Fig. 2.1.1.1.1. During operation, the wire loop is excited with a signal of frequency ranging from 10 to 50 kHz. When a vehicle (or any metallic object) stops or passes over the loop, the inductance of the loop is reduced and a change in oscillator frequency is induced. If this change in frequency exceeds a pre-defined threshold, a signal will be sent to the controller indicating the detection of a vehicle. A speed estimate is obtained by using a loop pair or using a single loop with some statistical algorithm [2.3, 2.4]. Classification is supported with newer versions of detector cards that can extract the raw inductive signature at a high sampling rate [section 5].

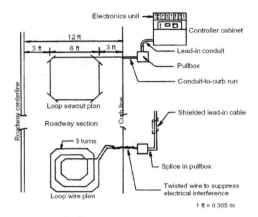

Fig. 2.1.1.1.1 Basic setup for an inductive loop detector [2.1]

Inductive loop is already a mature technology. It is well recognized as the industrial standard because of its high detection accuracy (e.g. >97%) [2.2]. However, its biggest disadvantage is that it causes serious traffic disruption during installation and repair. This makes the installation and maintenance cost very experience in term of traffic delay. The loop wire is also subjected to stresses of traffic and temperature, making its failure rate relatively high. Advanced algorithms were developed to identify bad detectors based on volume and occupancy measurements [2.5]. Nevertheless, broken detectors are seldom replaced because this is intrusive. Therefore, alternative detectors that can give the same accuracy level with minimum traffic disruption are being actively researched, which is also the main motivation of this research project.

Pneumatic Tube

Pneumatic tube is installed by sticking a long rubber tube on the pavement surface, perpendicular to the traffic flow direction. When a vehicle's wheels pass over the pneumatic tube, a pulse of air pressure is transferred along the tube. An electrical signal is triggered to represent the detection of an axle (vehicle) when the pulse of air pressure closes an air switch [2.6]. Because of its quick installation and low power usage, it is commonly used for short-term study of traffic counting and classification by axle count and spacing. Because of its simple hardware configuration, the installation and maintenance cost are relatively low [section 2.1.2]. Drawbacks include inaccurate axle counting when truck and bus volumes are high; the sensitivity of the air switch (for detection) is temperature dependent; and the unavoidable wear off of the rubber tube requires frequent maintenances. Therefore, pneumatic tube is seldom used for long-term surveillance.

Piezoelectric Sensor

Similar to inductive loop, piezoelectric sensor is installed by embedding it under the pavement. It is constructed by a specially processed material (quartz) that will generate a voltage when subjected to mechanical impact or vibration. The magnitude such a voltage is proportional to the force or weight of the vehicle. Since the voltage is only generated when the applied force is changing, the measurement will decay to zero if the vehicle stays on the sensor [2.6]. Besides vehicle detection, classification is done by axle count, spacing and weight. Because of its capability of weight estimation, it is commonly used as part of a weigh-in-motion system. Its drawbacks are similar to those of inductive loop, including disruption of traffic for installation and repair, failures caused by traffic stress and resurfacing, and sensitivity dependence on temperature and vehicle speed. An example of piezoelectric sensor setup (LINEAS) [2.7] is shown in Fig. 2.1.1.1.2.

Fig. 2.1.1.1.2 Example of piezoelectric sensor setup (LINEAS) [2.7]

Weigh-In-Motion (WIM) System

WIM system is used to estimate a vehicle's gross weight when its wheels pass over the sensors [2.6, 2.8]. It is used to increase the capacity of a station that monitors the weight of trucks on a freeway. Such a weight control is important because overweight trucks deteriorate pavements. It can also be used for vehicle detection and classification by number of axles and spacing.

The primary WIM technologies are piezoelectric, bending plate, load cell, capacitance mat and fiber optic. Their mechanisms operate as follows:

i, The mechanism of piezoelectric sensor is studied in last section.

ii, The bending plate has strain gauges attached underside, that generates a signal proportional to the deflection of the plate when it is under a load. The dynamic load of the vehicle, as well as the static load is estimated from this signal and the calibration parameters.

14

iii, The load cell sensor contains a small amount of hydraulic fluid that causes a pressure transducer to generate a signal proportional to the load. It is one of the most accurate but also the most expensive WIM system [2.6].

iv, Capacitance mat is made by two or more metal plates that act as capacitor terminals. The distance between these plates decreases when a vehicle passes over the mat, inducing an increase in capacitance. This also alters the resonant frequency of the mat, and the change is transformed into a signal proportional to the axle weight.

v, The fiber optic WIM sensor is installed by sticking a thin tube on the pavement's surface. When a vehicle's wheel passes over the tube, the optical fibers are perturbed (e.g. bends, micro-bends, change in refractive index and dimension). These perturbations are measured by intrinsic or extrinsic sensing devices [2.2, 2.9]. It is getting more and more popular in WIM applications because of its low cost, high accuracy, and immunity from electromagnetic interference.

2.1.1.2 Non-Intrusive Technologies

Non-intrusive technologies refer to those that do not need any installation on or under the pavement, so that the installation and repair of such a system can be done without disrupting the traffic. The detectors are usually setup on the roadside, or at an overhead position. Examples of this type of technology include microwave radar, infrared, Video Image Processing (VIP), ultrasonic and passive acoustic array.

Microwave Radar

Radar – an acronym for RAdio Detection And Ranging [2.10], is a system that uses radio waves to detect, determine the direction, distance and speed of some target objects. Microwave refers to a wavelength between 1 to 30 cm and corresponding frequency 1 to 30 GHz. A typical setup for a microwave radar system is shown in Fig. 2.1.1.2.1

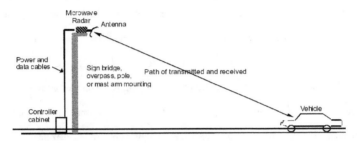

Fig. 2.1.1.2.1 Setup for a microwave radar system [2.6]

There are two types of microwave radar:

i, Continuous Wave (CW) Doppler radar [2.11] transmits a signal with constant frequency. When a vehicle passes the detection zone, a shift in the frequency is induced in the reflected signal (Doppler Effect). The detection and speed estimate of this moving vehicle can be measured from such a frequency shift. However, this type of radar cannot detect motionless vehicles.

ii, Frequency-Modulated Continuous Wave (FMCW) radar transmits a signal with constantly changing frequency. The time difference in transmitting and receiving a signal is used to determine the distance between the receiver and the target vehicle, as well as determining its present. Motionless vehicle can be detected. However, a pair of detection zones is needed for a speed estimate.

The main advantage of microwave radar is that the system performance is not affected by any weather change. The drawback is that CW Doppler radar cannot detect motionless vehicle unless an auxiliary device is equipped [2.6].

Infrared-Based System

Infrared (IR) radiation is a class of electromagnetic radiation with wavelength longer than that of visible light but shorter than that of radio waves. Common systems for traffic surveillance use IR ranging from 100 to 105 GHz. There are two types of IR-based system, active and passive:

i, Active IR system emits low-energy radiation by light-emitting diodes or high-energy one by laser diodes. The time difference between transmit and receive of the reflected signal from the detection zone is measured. A shorter return time represent the presence of a vehicle. Speed estimate is obtained by transmitting two or more IR signals onto different positions in the detection zone. Fig. 2.1.1.2.2 shows a simple setup for such a system.

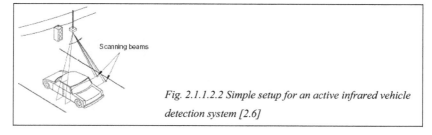

Fig. 2.1.1.2.2 Simple setup for an active infrared vehicle detection system [2.6]

ii, Passive IR system relies on the radiation emitted from vehicles and road surfaces (Gray body emission). In fact, any object with a temperature higher than the absolute zero (-273.15°C) emits radiation in the far IR part of the electromagnetic spectrum depending on the object's surface temperature, size and structure. Non-imaging systems use one or several energy-sensitive elements on a focal plane that gather energy from the detection zone. Imaging systems (e.g. Charge-Coupled Device (CCD) cameras), use two-dimensional arrays of energy-sensitive elements to reconstruct the pixel-resolution details from the imaged area [2.6]. Vehicles in the detection zone are detected by monitoring the change in the IR radiation received. The magnitude of signal from a target vehicle is proportional to the product of an emissivity difference term (between the road and the vehicle), and a temperature difference term (between the road surface and the atmosphere). Fig. 2.1.1.2.3 shows the pictures of two side-mounted passive IR systems.

Fig. 2.1.1.2.3 Pictures of two passive IR systems: IR 254 by ASIM Technologies Ltd. [2.12] and PIR-1 by Siemens Energy and Automation, Inc. [2.13]

Besides traffic counts, multi-channel and multi-zone IR systems provide speed estimates, as well as vehicle lengths for classification. The main advantage of an IR system is its feasibility of transmitting multiple beams for multi-zone detection in a single detector unit. The drawback is that its performance is greatly affected by the environment: confusing signal from sunlight, IR energy is absorbed or scattered by atmospheric particulates, fog, rain and snow [2.14].

18

Video Image Processing (VIP)

A VIP system includes one or several video cameras, microprocessor-based equipment for digitizing and processing the imagery, computer and software for analyzing the images to extract traffic data. In general, vehicle detection is done by monitoring the changes between successive video frames. A simple approach is to analyze the variations in the gray levels of the black-and-white pixel groups induced by vehicles passing the detection zone [2.6].

There are three types of VIP systems: tripline, closed-loop tracking and data association tracking [2.16, 2.6].

i, Tripline systems monitor changes in pixels caused by a vehicle relative to an empty detection zone. Images are analyzed by surface-based or grid-based algorithms, which identify edge features of vehicle or classify squares on a fixed grid into moving, stopped or no vehicle respectively.

ii, Closed-loop tracking systems continuously track vehicles through the camera's field of view, by validating multiple detections of the same vehicle along a track.

iii, Data association tracking systems track a particular vehicle or group of vehicles by extracting connected areas of pixels.

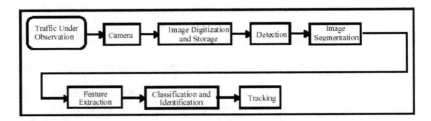

Fig. 2.1.1.2.4 Flow diagram of a typical VIP system for vehicle detection, classification and tracking [2.15]

The flow diagram of a typical VIP system is shown in Fig. 2.1.1.2.4. Images captured by the cameras are usually digitized by a microprocessor card and stored into a computer. Vehicle detections are conducted on a series of images. Image segmentation is used to divide the image area into smaller regions where features can be better extracted. The extracted features are used for classification and tracking. With the tracking results, vehicle trajectories of identified vehicles can be obtained, which can be used to provide lane changing and origins/destinations statistics.

It is known that the performance of a VIP system is affect by many environmental factors, such as lighting condition (daylight or vehicle headlight), shadow and snow. Many different image processing algorithms are proposed to improve and maintain accuracy level under non-ideal environmental conditions. One of the popular approaches is using artificial neural network [2.17]. The detection accuracy of a modern VIP system is high. Combined results for clear and inclement weather show vehicle detection and speed estimate accuracies of a correctly calibrated VIP system is greater than 95% [2.18].

20

The disadvantages of VIP systems include performance greatly affected by inclement weather; false detection caused by vehicle's shadows projected onto adjacent lanes; camera vibration caused by strong wind; and the requirement of a high mounting setup for the cameras (up to 60 feet height). The installation and equipment cost is relative high [section 2.1.2] and the system is only cost effective if many detection zones are required within the field of view of the camera.

Ultrasonic System

Ultrasonic refers to those high frequency sound waves which are beyond a human's audible range; waves of frequency between 25 and 50 kHz are commonly used. Its principle mechanism is similar to that of microwave radar. Sound pulses are transmitted and the reflected pulses are received, and the distance from the receiver to the road or vehicle surfaces is measured according to the wave travel time. If a distance smaller than that to the background road surface is measured, the presence of a vehicle is declared. Speed estimate is obtained by deploying multiple detection zones.

Fig. 2.1.1.2.5 Picture of a typical ultrasonic system setup: Lane King by NOVAX Industries Corp [2.19].

The picture of a typical ultrasonic system is shown in Fig. 2.1.1.2.5. Constant frequency ultrasonic systems that measure speed using Doppler principle are also available on the market. However, they are much more expensive than the pulse models and therefore rarely used. The disadvantage of ultrasonic system is that its performance is affected by

21

temperature change and air turbulence. Some modern models do have temperature compensation built in.

Passive Acoustic System

Passive acoustic systems measure the acoustic energy or audible sounds produced by a variety of sources within a vehicle. The overall sound energy level increases when a vehicle passes the detection zone. Besides vehicle detection and speed estimate by a pair of detection zones, classification can be done by applying pattern matching and neural network on the acoustic signatures [2.20].

Fig. 2.1.1.2.6 Picture of a multi-lane passive acoustic system: SmarTek SAS-1 [2.21]

The picture of a multi-lane passive acoustic system, SmarTek SAS-1 [2.21], is shown in Fig. 2.1.1.2.6. It uses a fully populated microphone array and adaptive spatial processing to form multiple detection zones. It can monitor up to 7 lanes if the device is mounted over the center of the roadway, and 5 lanes when mounted on the side. This multi-lane design is a great advantage for highway deployment as 5-7 dual loops setup can be replaced by a single device. Drawbacks include performance affected by temperature and detection accuracy drops with slow moving vehicles.

2.1.1.3 Off-Roadway Technologies

Off-Roadway Technologies refer to those that do not need any hardware to be setup under the pavement or on the roadside. It includes probe vehicle technologies with Global Positioning System (GPS) and mobile phones; Automatic Vehicle Identification (AVI); and remote sensing technologies that make use of images from aircraft or satellite [2.2].

Probe Vehicles with Global Positioning System (GPS)

GPS is a satellite navigation system originally developed by the United States Department of Defense, officially named NAVSTAR GPS in 1978 [2.22]. A constellation of more than 24 GPS satellites broadcasts precise timing radio signals to GPS receivers. The location and speed is calculated by the multilateration technique that accurately computes the time difference of arrival (TDOA) of a signal transmitted from three or more synchronized transmitters. The system is available for free uses in civilian application as a public good.

For traffic surveillance, probe vehicles equipped with GPS receivers are driven through the traffic sections of interest. Their position and speed information determined from the GPS is transmitted back to the Traffic Management Center (TMC) for travel time and section speed analysis [2.23]. A sample configuration of a GPS-based probe vehicle system is shown in Fig. 2.1.1.3.1 [6.6]. Drawbacks include lack of point traffic statistics at a fix location, and the fact that system coverage is limited by the number of probe vehicles.

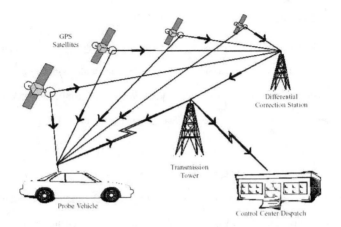

Fig. 2.1.1.3.1 Sample configuration of a GPS-based probe vehicle system [6.6]

Probe Vehicles with Mobile Phones

The localization technique [2.30] is similar to that of a GPS system, with the satellites

replaced by phone antenna base stations, and GPS receivers replaced by mobile phones.

Because of the high penetration rate of mobile phones, at least one mobile phone can be

found in a traveling vehicle. For traffic surveillance, either active reporting by volunteer

drivers or passive mobile phones localization can be applied. Depending on the density of

mobile phone antenna stations, the accuracy of such a localization technique can be as good

as a hundred meters in urban areas, but as poor as 30 km in suburban areas [2.24]. A high

percentage of coverage on main arterials can be achieved if this system is deployed in a

national scale. Unfortunately, privacy concern is raised by the public on the unauthorized

use of one's location information which makes it not suitable for large scale deployment.

Automatic Vehicle Identification (AVI)

AVI refers to the technology that use roadside antennae to read the identification number of transponders equipped on probe vehicles. The section travel time can be determined if the probe vehicle travel through more than one antenna station. This technology is primarily used in electronic toll collection. However, the limited number of AVI antenna stations restricts data collection capability, so the system is usually used for long range travel time estimate only [6.6].

Remote Sensing

Remote sensing refers to the technologies that collect traffic information without direct communication or physical contract with the vehicles or roads. Basically, high-resolution imagery from aircraft or satellite is used to extract traffic information like traffic count and speed. In [2.25], a satellite was used to monitor a traffic network and the collected data were used to improve the Annual Average Daily Traffic (AADT) accuracy. Again, the system coverage is limited by the availability of the aircrafts and satellites.

2.1.2 Comparison of Different Surveillance Technologies

In this section, different surveillance technologies are compared in terms of their data type availability, system performance and system cost. The comparison is based on the results of many experimental evaluation cases summarized in [2.2, 2.6, 2.14]. The extracted results are presented as below:

Data Type

Technology	Data Type				
	Count	Speed	Classification	Occupancy	Presence
Intrusive					
Inductive Loop	Y	Y	Y	Y	Y
pneumatic road tube	Y	Y	Y	N	N
piezoelectric cable	Y	Y	Y	N	N
Non-Intrusive					
WIM system	Y	Y	Y	N	N
Microwave Radar					
CW Doppler	Y	Y	Y	Y	N
FMCW	Y	Y	Y	Y	Y
Infrared					
Active	Y	Y	Y	N	N
Passive	Y	Y	Y	Y	Y
Video Image Processing	Y	Y	Y	Y	Y
Ultrasonic	Y	N	N	N	Y
Passive Acoustic	Y	Y	Y	Y	Y
Wireless Sensor Network					
Magnetometer	Y	Y	Y	Y	Y

Y: available, N: not available

Table 2.1.2.1 Data type available in different surveillance technologies [2.2]

Count, speed, classification, occupancy and presence are the basic data types obtained from traffic surveillance. Table 2.1.2.1 [2.2] shows the availability of these data types in different technologies. Traffic count is available in all the technologies studied. Speed measurement usually requires a dual-detection-zone configuration with synchronized time and fixed separation. For system with a single detection zone (i.e. a single inductive loop), a rough speed estimate is obtained by assuming the vehicle length to be a fixed value. Doppler-based technology can be used to provide speed estimate with a single sensor. However, it cannot provide presence data as it does not response to motionless vehicles. Vehicle type classification data is usually obtained by analyzing the detected vehicle lengths, heights, number of axles and spacing [section 5.1]. Other data types like section travel time and origin/destination matrix are not directly available in most detection

systems. Remote sensing and reidentification systems [Ch. 6] are used to obtain this type of

information.

System Performance

System		Mounting	Error [%]		Sources
			Count	Speed	
Inductive Loop					
	Saw-cut	Pavement	0.1-3	1.2-3.3	MNDOT[2.26]
Pneumatic Road tube					
		Pavement	0.92-30		SDDOT[2.27]
Microwave Radar					
	TDN 30	Overhead	2.5-13.8	1	MNDOT[2.28]
	RTMS	Overhead	2	7.9	MNDOT[2.28]
Active Infrared					
	Autosense II	Overhead	0.7	5.8	MNDOT[2.26]
Passive Infrared					
	ASIM IR 254	Overhead	10	10.8	MNDOT[2.26]
Video Image Processing					
	Autoscope solo	Side-fire	5	8	MNDOT[2.26]
	Autoscope solo	Overhead	5	2.5-7	MNDOT[2.26]
Ultrasonic					
	Lane King	Overhead	1.2		MNDOT[2.28]
Passive Acoustic					
	SAS-I	Side-fire	8-16	4.8-6.3	MNDOT[2.26]
Wireless Sensor Networks					
	VSN240	Pavement	1-3		[section 4.4]

Table 2.1.2.2 Error rate of different surveillance technologies in field tests [2.2]

System performance statistics of surveillance products provided by the vendors are usually

exaggerated, as they trend to use ideal conditions for the evaluation. On the other hand, the

real-life performance of these technologies was studied in many academic researches [2.26,

2.27, 2.28]. Among these field tests conducted under real world environment, some of the

count and speed accuracy results were extracted from [2.2] and presented in Table 2.1.2.2.

Inductive loop detector is one of the most accurate count detectors. In [2.26], it gave an error rate of 0.1-3% for counting vehicles in a one-hour period on the freeway. The corresponding speed difference between the loop data and probe vehicle data was 1.2-3.3%. In section 4.4, experimental field tests show that the proposed wireless sensor networks give an counting error rate of 1-3%, which is closely compatible to that of inductive loop detector.

The system performance may change under the influence of uncontrollable environmental conditions. As noted in section 2.1.1, different technologies are affected by different environmental conditions. A summary of environmental factors that affect the performance of different surveillance technologies is shown in Table 2.1.2.3 [2.2].

Technology	Environmental Factor				
	Penetration	Wind	Temperature	Lighting	High traffic flow
Intrusive					
Inductive Loop			Y		
Pneumatic Road Tube			Y		Y
Piezoelectric Cable			Y		
Non-Intrusive					
Microwave Radar					
CW Doppler					Y
FMCW					
Infrared					
Active	Y				
Passive					
Video Image Processing	Y	Y	Y	Y	
Ultrasonic					
Passive Acoustic	Y		Y		Y
Wireless Sensor Network					
Magnetometer			Y		

Y: Affected

Table 2.1.2.3 Environmental factors that affect the performance of different surveillance technologies [2.2]

System Cost

A cost comparison between different surveillance technologies should include device cost,
installation and maintenance cost. The system cost depends on the configurations and
requirements of specific application. In this analysis, a typical vehicle count and speed
estimate application deployed on freeway is considered: monitoring three lanes in each of
the traveling directions. For example, two inductive loops are placed in each of the six
lanes, the device and installation cost is 12x750=$9000. The annualized life-cycle cost is
calculated according to Eq. 2.1.2.1. Taking data sources from [2.2], the life-cycle costs of
different surveillance systems are presented in Table 2.1.2.4.

$$LifeCycleCost = ((DeviceCost * Quantity) + InstallationCost)\left[\frac{i(1+i)^{oy}}{(1+i)^{oy}-1}\right] + AnnualMaintenanceCost$$

OY = System lifetime in year
i = interest rate (0.04 is used)   (Eq. 2.1.2.1)

Technology	Device Cost [$]	Installation Cost [$]	Maintenance Cost [$ / yr]	Lifetime [yr]	Life-Cycle Cost [$]
Inductive Loop					
Saw-cut	12x750=9000	<- Included	700	10	1810
Microwave Radar					
TDN 30	6x995=5970	3200	600	7	2130
RTMS	2x3300=6600	400	200	7	1370
Active Infrared					
Autosense II	6x6000=36000	3200	600	7	7130
Passive Infrared					
ASIM IR 254	6x700=4200	1200	600	7	1500
Video Image Processing					
Autoscope solo	2x4900=9800	1000	400	10	1730
Ultrasonic					
TC 30	2x735=1470	400	200	7	510
Passive Acoustic					
SAS-I	2x3500=7000	800	400	7	1700

29

Wireless Sensor Networks					
VSN240	450x12=5400	200	200	10	890

Table 2.1.2.4 Estimated life-cycle costs of a typical freeway application [2.2]

With this typical freeway application, the life-cycle cost of a set of 12 inductive loops is $1810. If a set of 12 VSN240 sensor nodes is used to replace the system, the life-cycle cost can be cut by half and drop down to $890. This life-cycle cost analysis does not include the traffic delay cost caused by disrupting the traffic during installation and maintenance. Details on the motivation for using such a wireless sensor networks for traffic surveillance are presented in the next section.

2.2 Motivation for Using Wireless Sensor Networks

The increasing traffic congestion is a growing problem in many countries. The 2005 Urban Mobility Report [1.1] shows that the total cost of congestion for 85 U.S. urban areas is estimated to be 65 billion dollars per year, which come from 3.5 billion hours of delay and 5.7 billion gallons of excess fuel consumed. Besides building new roads and bridges to ease congestion, Intelligent Transportation Systems (ITS) seek to maximize the capacity of existing traffic networks and minimize the associated delay.

Accurate and reliable real-time traffic data from surveillance systems is essential for the efficient and successful execution of all ITS systems. For example, traveler information system, freeway and arterial management systems, emergency management and parking management rely on the coverage and accuracy of the real-time traffic information [2.31]. In order to maximize the benefits from all these ITS technologies, a large scale deployment of traffic controls on all major freeways and local streets must be under taken. Therefore, real-time traffic information at all these sites is required. This presents a serious challenge to the development of surveillance industry.

Because of the highly intrusive characteristic of inductive loop detectors, the quest for researching a reliable and cost-effective alternative system, which can provide traffic data at the same accuracy level as inductive loop systems, while minimizing the disruption during installation and maintenance, has been underway for some time. The motivation of developing wireless sensor networks based surveillance system is to provide a direct replacement for the inductive loop systems, and extend the coverage of ITS applications over all the freeways and local intersections. Such a large scale deployment has the potential to revolutionize the whole traffic surveillance and control industry.

Flexibility

Wireless sensor networks have a high level of flexibility in their deployment configuration. Since the sensor nodes can be placed virtually anywhere on the road as long as they are within communication range, customized configurations can be adopted for different applications and environments. This unique characteristic is a big advantage over all other surveillance technologies.

Multi-Functional

A multi-functions wireless surveillance system [section 7.2] can be developed by adding other sensing modalities to the existing sensor node platforms. Temperature sensors can be added to detect ice and snow; humidity sensors can be added to detect rain and fog; accelerometers can be added to monitor structures of bridge and pavement. This multi-functional characteristic further extends the possibility of more advanced ITS applications.

Wireless Communication Capability

Research on safety control by inter-vehicle communication (IVC) and road-to-vehicle communication (RVC) [2.32] is being actively conducted. The sensor nodes can be used to extend the communication networks of IVC and RVC by simply using the standard protocol, IEEE 802.11p [2.33] and Dedicated Short Range Communications (DSRC) [7.28]. This feature is extremely useful in enhancing the safety control at intersections, where traffic lights and warning signs can be controlled in advance.

Besides all these valuable characteristics, prototypes of the wireless sensor networks also demonstrate that its detection accuracy [section 4.4] is as high as that of inductive loop detectors. Vehicle classification [Ch. 5] and reidentification [Ch. 6] can also be achieved using the same hardware platform. These promising results give us a strong reason for investing more resources on the research and development of wireless sensor networks for traffic surveillance.

Ch. 3 Wireless Sensor Networks

A Wireless Sensor Network (WSN) [3.1] is a network of small sensor nodes (SN) communicating among themselves using wireless communication, to sense the physical world. It combines distributed sensing, computation and wireless communication technologies. Conditions such as temperature, sound, vibration, pressure, motion or pollutants could be monitored on a large scale using a spatially distributed WSN (from tens to thousands of nodes). Because of its variety in function and flexibility in deployment, numerous potential applications could be developed using WSN.

WSN has gained a significant amount of public attention in recent years. In MIT's Technology Review magazine in 2003 [3.2], WSN was picked as one the ten technologies that will change the world. Thanks to the revolution in sensor, processor, communication and power technologies, sensor nodes can now be integrated into a small millimeter-cubic size at low cost [3.3]. Such technology advances push WSN into a new era as it is now flexible and cost-effective to be deployed on a large scale.

In this chapter, we will discuss the architecture and components of a WSN, how it could be used in the traffic surveillance industry, and the corresponding hardware and software specifications of the prototypes that were developed, as well as the communication protocols and lifetime analysis.

3.1 Architecture and Components

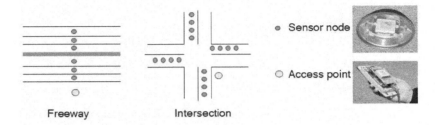

Fig 3.1.1 A sample wireless sensor network layout for traffic surveillance

The physical architecture of a wireless sensor network consists of a number of Sensor Nodes (SN) and a single Access Point (AP). The AP has superior computation resources, enhanced radio communication and unlimited power supply. In general, some environmental conditions are measured by the sensor nodes deployed with a spatial density and at a sampling rate specified by the application. The raw sensor signals are first processed by the processor in the sensor node to extract some useful information. The output of this initial sensor node-based processing is transmitted to the access point either through a direct communication with the access point or a multi-hop communication across other sensor nodes. Eventually, the access point processes the data collected from all the sensors in the network to extract more information, places this information into some meaningful format and sends it to the end user or some other control system.

In the traffic surveillance application, magnetic sensors placed at known locations on the pavement obtain the magnetic 'signature' of vehicles traveling over the sensors. The signals are processed through a vehicle detection algorithm by the sensor nodes, detection events are then generated and transmitted to the access point. Having collected the event data from these synchronized sensor nodes; the access point can then calculate the counts, occupancy and speed of the monitored traffic. Finally, this real time traffic information is passed to the Traffic Management Center (TMC) or to the local control unit for applications like traffic monitoring and signal control. Details of the vehicle detection algorithm are provided in Ch. 4.

A sensor node is comprised of four components: sensors, processor (microcontroller), radio, and power source. Some features of these components for the sensor networks are discussed in the following section.

3.1.1 Sensor

A sensor is a transducer that transforms some physical process into an electrical signal, which can be measured by a digital processor. Many sensors can provide information of interest for traffic surveillance, such as temperature, humidity, pollutant, vibration (accelerometer), photonic (lighting condition), acoustic and magnetic sensors. It is possible to provide a detailed picture of the road conditions and traffic flow with a combination of these sensors.

Thanks to advances in MEMS (Micro-Electro-Mechanical System) technology [3.4], it is now possible to integrate many sensors into a single, small integrated circuit board (the size of a quarter) with a very low power consumption and at a relative low cost. The power consumption of the sensor could be as low as 1mA@3V [3.5] with 0.1% duty cycle, which

contributes a negligible fraction of the total power consumption A lifetime analysis of the

system is presented in section 3.3.3. These new generation of MEMS sensors have opened

the door to many sensor network applications that were not feasible in the past. For

example, WSN have been developed for environmental applications like habitat

monitoring, animal tracking and forest-fire detection [3.6, 3.7].

3.1.2 Processor (microcontroller)

A microcontroller is a computer-on-a-chip used to control electronic devices. It is a type of

microprocessor emphasizing self-sufficiency and cost-effectiveness, in contrast to a

general-purpose microprocessor. All the processing units, memory, analog-to-digital

converter, digital I/O interface and peripherals are integrated onto a single integrated

circuit. This integration drastically reduces the number of chips, the amount of wiring and

PCB space that is needed.

Most micro-controllers used on a sensor node are Digital Signal Processors (DSPs) [3.8].

They are responsible for sampling and converting the analog electric signal from the

sensors into a digital format, and carry out the necessary data processing and radio

communication control.

Sleep Mode

An important feature of a microcontroller for sensor networks is its sleep mode

characteristic. Since the microcontroller is expected to be in sleep mode (idle) 99.9% of the time, its sleep mode power consumption and wake-up time become critical parameters of the system. Sleep mode current consumption varies between 1 uA and 50 uA [3.9], and the wake-up time varies between 6 us and 10ms across controller families. The faster the wake-up time, the quicker the controller can enter and leave its sleep mode, so that the sleep mode can be used at a higher frequency to increase the responsiveness of the sensor node, without a large increase in the power consumption.

3.1.3 Radio

The radio plays a critical role in the lifetime of a sensor node, because the overall power consumption is dominated by the energy cost of radio communication. Typically, more than 90% of the energy consumed in the node is accounted for by the radio transceiver [3.10]. The energy consumed in transmission and in reception are comparable in the low power radios that are commonly used in a sensor node. This implies that the duration of both transmission and reception are important concerns in designing a communication protocol, which should aim at minimizing the power consumption, thereby maximizing the sensor node lifetime. This topic is further discussed in the communication protocol and lifetime analysis section of this chapter.

The relationship between the received power and the transmission distance follows a power law with an exponent between -3 and -4 [3.11]. The communication range is also affected by the sensitivity of the receiver, and the gain and efficiency of antenna. Typical receiver sensitivities are between -85 and -110 dBm. When transmitting at 0 dBm, a receiver sensitivity of -85 dBm will result in an outdoor free space range of 25-50 meters. This

37

communication setting is an important design factor in determining how many nodes are needed to meet the application's need and maximizing system lifetime.

3.1.4 Power Source

The lifetime of a WSN directly depends on its power source. In fact, the energy constraint is a dominant factor of system design trade-offs for small embedded sensor devices. The scaling down in size and cost of microcontroller and sensor has outpaced that of a battery. This makes the power source account for a growing portion of the cost of a WSN. This cost is further magnified by the maintenance cost of replacing or recharging the batteries on a regular basis. This implies that no single power source will suffice for all applications, and the choice of a power source needs to be included in the system design.

Zinc-based batteries have high energy density but high leakage, so they are only good for use in short-duration applications. Lithium batteries offer higher energy density with fewer memory effects[1], but the drawback is that they require longer recharge times. The newly developed polymer-based batteries [3.13] have excellent energy density with flexible form factor, but they are relatively expensive (e.g. $20 / 1500mAh). Another developing battery technology is the micro fuel cell [3.12], which has 10 times larger energy density than the commonly used alkaline battery. Lastly, we note that there is active research aimed at collecting renewable energy from wind or solar panels. Table *3.1.4.1* below gives a summary of the energy ratings of different types of batteries [3.1].

---

[1] Memory effect refers to the limitation in recharging if the batter is not completely discharged first.

Battery types:	Alkaline	Zinc-air	Lithium	Fuel Cell	Solar (outdoor)
Units:	J/cm^3				mW/cm^2
Energy ratings:	1190	3780	2880	8900	15

Table 3.1.4.1 Energy rating of different power sources [3.1]

3.2 Hardware and Software Specifications

MICA2Dot

Protective cover
(SmartStud)

Magnetometer
(HMC 1051Z)

Battery (TL-5135)

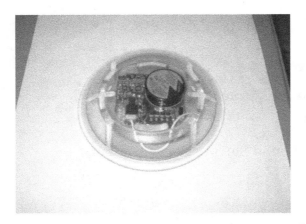

Fig. 3.2.1 Pictures of first generation prototype of the sensor node: TrafficDot

Fig. 3.2.1 above shows the first generation prototype of the sensor node, called TrafficDot. It was designed as a sensor node prototype for traffic surveillance research, and manually built with the assistance of Tom Oberheim. The basic components include: two magnetic sensors (HMC1051Z [3.14]), a microcontroller with integrated radio (MICA2DOT [3.15]), a battery (TL-5135 [3.16]) and a protective cover (SmartStud [3.17]) made of a high impact polycarbonate and designed to be placed on the road pavement.

Even though this first prototype does not have full functionality, it shows a very promising future in the application of wireless sensor networks on traffic surveillance. Moreover, it identified important directions for improving the design and development of the second and third generation prototypes, which eventually resulted in a commercial product [3.18]. Most of the data presented in this thesis were generated from these sensor node prototypes.

In this section, we will discuss their hardware and software specifications, and provide a picture of the experimental sensor network that was developed.

3.2.1 Magnetic Sensor

Since almost all vehicles have significant amounts of ferrous metals in their chassis (iron, steel, nickel, cobalt, etc.), the magnetic field disturbance created by a vehicle is sufficient to be detected by a magnetic sensor, which makes it a good candidate for detecting vehicles[2]. Potential vehicle detection applications that could use these magnetic sensors include traffic surveillance, railroad crossing control, parking lot space monitoring and automatic gate opening.

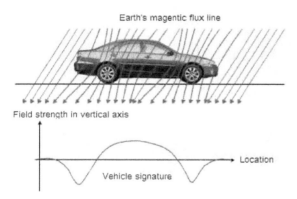

Fig 3.2.1.1 The disturbance of Earth's magnetic flux lines by a vehicle

Fig. 3.2.1.1 gives a pictorial representation of the disturbance of the magnetic flux lines when the Earth's magnetic field penetrates a vehicle. The earth's magnetic field strength is

---

[2] Magnetic sensors will not detect bicycles made of composite material or pedestrians.

41

roughly equal to a half-gauss in magnetic flux density [3.19], so a low field and high sensitivity magnetic sensor is needed to measure the disturbance. One type of these low field magnetic sensors is the magneto-resistive sensor, which is very suitable for use in a sensor node because of its small size. Magneto-resistive sensor can be further classified into Anisotropic Magneto-Resistive (AMR) and Giant Magneto-Resistive (GMR) types. An AMR sensor is directional, which means it only provides an amplitude response to the magnetic field along its sensitive axis, whereas a GMR sensor has little directionality.

AMR Sensors

For traffic surveillance, it is essential to isolate one vehicle's magnetic signal from the signals from vehicles in different lanes and travel direction. So an AMR sensor is a much better choice for use in the sensor node. Basically, the AMR sensor is a Wheatstone bridge device as shown in Fig 3.2.1.2 [3.19]. It is made out of a nickel-iron (Permalloy) thin-film deposited on a silicon wafer and patterned as a resistive strip element. In the presence of a magnetic field, a change in the bridge resistive elements causes a corresponding change in voltage across the bridge outputs. These resistive elements are aligned together to have a common sensitive axis that will provide positive voltage change with magnetic fields increasing in the sensitive direction.

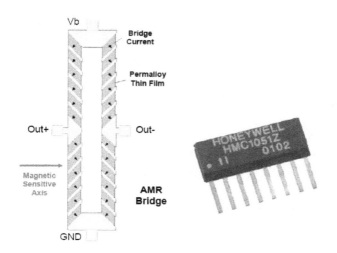

Fig 3.2.1.2 AMR Sensor Bridge   *Fig 3.2.1.3 Honeywell HMC1051Z magnetic sensor*

Honeywell HMC1051Z

The AMR sensor Honeywell HMC1051Z [3.14] was used in the sensor node prototype

because of its high sensitivity (1.0 mV/V/gauss), small size (10x4x1.5 mm) and low power

consumption (5mA when sampling).

Characteristics	Min	Typical	Max	Units
Operating temperature	-40		125	°C
Field Range	-6		+6	gauss
Sensitivity	0.8	1.0	1.2	mV/V/gauss
Resolution		120		μ gauss
Bandwidth		5		MHz

Table 3.2.1.1 Summary of characteristics of HMC1051Z

One drawback of HMC1051Z is that it requires a high set/reset current (0.1 ~ 0.5A) [3.14] to be driven through the bridge periodically, in order to maintain its high sensitivity. This not only increases the power consumption, but also puts limits on the choice of a battery since most low power batteries are not designed to deliver such a large current. More detailed analysis of the magnetic signal for vehicle detection is provided in Ch. 4.

3.2.2 MICA2DOT (MRP510CA)

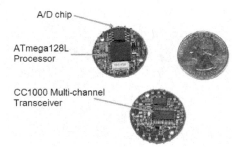

Fig 3.2.2.1 Quarter-size Crossbow MICA2DOT (MPR510CA)

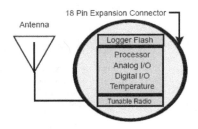

Fig 3.2.2.2 Block diagram of Crossbow MICA2DOT (MPR510CA)

MICA2DOT (MPR510CA) [3.15] is a wireless sensor networks product manufactured by Crossbow Technology, and originally designed and developed under the TinyOS project [3.20] in UC Berkeley. It was designed to be a sensor node deployment platform for embedded wireless sensor networks. Being the third generation of its family, it has all the necessary components (processor, I/O interface, radio and memory) integrated into a quarter-sized (25mm) form factor. Because of its small size, it is also known as "Mote". Its low-power processor (Atmel ATmega128L [3.23]), low-power radio (Chipcon CC1000 [3.24]) and flexible operating system (TinyOS [3.20]) allows this base platform to be easily connected to other compatible sensor or data acquisition modules. Moreover, a base station could be setup quickly by connecting a MICA2DOT to a PC, and forward the communication data to any desired program or interface. As a result, it was used in our sensor node prototype design. A summary of its characteristics is given in the following table.

Characteristics	Value	Remarks
Processor Performance		
Program Flash Memory	128K bytes	
Measurement (Serial) Flash	512K bytes	>100,000 Measurements
Serial Communications	UART	0-3V transmission levels
Analog to Digital Converter	10 bit ADC	6 channel, 0-3V input
Other Interfaces	DIO	9 channels
Current Draw	8 mA	Active mode
	< 15 µA	Sleep mode
Multi-Channel Radio		
Center Frequency	433 MHz	ISM bands
Data Rate	38.4 Kbaud	Manchester encoded

RF Power	-20 to + 10 dBm	Programmable, typical
Receive Sensitivity	-101 dBm	Typical, analog RSSI
Outdoor Range	1000 ft	1/4 Wave dipole, line of sight
Current Draw	25 mA	Transmit with maximum power
	8 mA	Receive
	< 1 µA	Sleep

Table 3.2.2.1 Summary of characteristics of MICA2DOT (MPR510CA)

TinyOS

MICA2DOT can be programmed using TinyOS [3.20], which is an open-source operating system designed for wireless sensor networks implementation. It features a component-based architecture such that only a minimal amount of modification of the main programming codes is needed when switching between hardware components (e.g. different sensors, communication protocols, etc.). This key feature enables rapid innovation and implementation of different applications while minimizing code size as required by the severe memory constraints inherent in the sensor nodes. With the open-source component library for TinyOS, all the programming codes for sensor drivers, data acquisition tools and network protocols could be further refined to suit a custom application's configuration.

Another key feature of TinyOS is its event-driven execution architecture. It allows a sensor node to remain in the sleep mode until an event is trigged by the physical world interface or the processor itself. This feature makes the sensor network system more flexible in handling the sensor interface and in scheduling efficient communication. As a result, fine-grained power management can be implemented according to the application's system properties.

In addition to the Crossbow Sensor Motes family, TinyOS has been ported to over a dozen platforms and many sensor boards. Open-source libraries are also available for simulating the TinyOS system in a complete sensor network. A wide community uses it to develop and test various algorithms and protocols of sensor network communication. Lastly, because of its open-source nature, several groups are actively contributing code to the sourceforge site, making TinyOS more stable while expanding its functionality.

3.2.3 Battery

It is an important task to choose a suitable battery that will fit the sensor nodes and applications since its performance directly determines the lifetime of a sensor node, as well as the whole network system. It is even more critical for the TrafficDot prototype as it must be fitted into a small space limited by a special protective container, as well as providing reasonable power over its lifetime while periodically supplying the set/reset current (0.1 ~ 0.5A) [3.19] for the magnetic sensor HMC1051Z.

Lithium Battery

After researching the available battery products, it was found that a Lithium battery outperforms the commonly used Alkaline battery in every way. A Lithium battery not only produces three times more energy density than its alkaline counterpart [3.1], it also provides a more consistent output voltage and allows higher maximum current to be drawn. In view of this, Lithium battery finds application in many long-life, critical devices, such as cardiac pacemakers, which are designed to last more than 15 years. Below is a table of a comparison of their characteristics.

	Lithium	Alkaline
Energy Density	900 Wh/L (@3.3V)	320 Wh/L (@1.2V)
Operating Temperature	-55° C to +200° C	-20° C to +54° C
Nominal Voltage	3.9 V ~ 3.6 V	1.5 V
Watt Hours (D size cell)	59.0	22.5

Table 3.2.3.1 Comparison of characteristics of Lithium and Alkaline batteries

The term "lithium battery" actually refers to a family of different chemistries, comprising many types of cathodes and electrolytes. One type of lithium cell having a large energy density is the lithium-thionyl chloride cell. In this cell, a liquid mixture of thionyl chloride and lithium tetrachloroaluminate acts as the cathode and electrolyte respectively. It can provide extremely high currents and can discharge very rapidly when short-circuited. Although this is useful in applications where high currents are required, a too-rapid discharge of a lithium battery can result in overheating of the battery, or even in an explosion. Lithium-thionyl chloride batteries are especially capable of this type of discharge, so consumer batteries of this type usually incorporate over-current or thermal protection in order to prevent explosion.

TADIRAN Lithium Battery TL-5135

The Lithium Thionyl Chloride battery TL-5135 [3.16] manufactured by TADIRAN was chosen for the TrafficDot prototype. It was picked mainly because of its high power capacity (1.7Ah) and its stable discharge characteristic even when supplying power at a relative high current. Below is a summary of its characteristics and a plot of its discharge

voltage at different current levels.

Nominal capacity @ 0.5 mA	1.7 Ah
Rated voltage	3.6 V
Size	⌀32 x 10 mm
Weight	21.5g
Operating temperature	-55°C to +85°C

Table 3.2.3.2 Summary of characteristics of TADIRAN Lithium Battery TL-5135

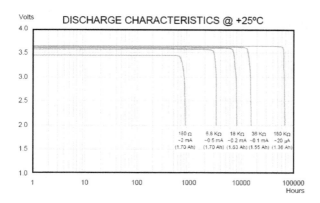

Fig 3.2.3.1 Discharge characteristic of TL-5135 at different current levels

3.2.4 SmartStud Container

Since the TrafficDot sensor node has to be placed on the pavement, a heavy duty container

is needed to protect it against the impact of vehicles. Instead of designing a new container,

we tried to adopt pavement markers available on the market and searched for one with

enough space to house all the sensor node components. These requirements were met by the SmartStud System [3.17] from a New Zealand based company that manufactures LED

pavement marker systems.

Fig 3.2.4.1 Pictures of SmartStud containers

The SmartStud container was originally designed to house a circuit for flashing LED on a

pavement. The design left enough space for fitting all the sensor node components inside.

The shell material is a Bayer high impact polycarbonate that can withstand 9 kN of weight

so as to satisfy the industrial standards for placing it on the road.

3.2.5 Sensys Networks, Inc.

The first generation TrafficDot prototype was tested in controlled areas such as the

Richmond Field Station (RFS) of PATH, as well as in urban traffic around the campus.

Even through a loss in data communication was experienced, caused by the attenuation of

the radio signal from some moving vehicles, it demonstrated the potential and promising

future of deploying such a wireless sensor networks for traffic surveillance. Prof. Pravin

Varaiya found it an opportune time to commercialize this product and co-founded a

company with Amine Haoui and Robert Kavaler called Sensys Networks, Inc. [3.18]. As a result, new prototypes of sensor node and access point were designed and developed in a more professional approach.

Fig 3.2.5.1 Second prototype generation of TrafficDot

Fig 3.2.5.2 Third prototype generation of TrafficDot

(VSN240-s)   (VSN240-f)   (AP240-e)   (RP240-b)

Fig 3.2.5.3 VSN240 family products manufactured by Sensys Networks

The Sensys Networks' VSN240 family shown above is a family of sensor nodes (SN) and access point (AP) designed specifically for use in traffic surveillance. VSN240-s is the commercial version of TrafficDot with three-axis magnetic sensors HMC1051Z. VSN240-f is a flush-mount counterpart of VSN240-s which can be installed under the pavement. AP240-e is the access point that can provide a one-hop communication with the sensor nodes within a range of 50m, and up to 96 sensor nodes can communicate with it on a single radio channel. Moreover, the collected data can be transferred through an ethernet port or wirelessly using GPRS [3.21] or CDMA [3.22]. RP240-b Repeater is a standalone, solar or line-powered device that extends the radius of coverage of a VDS240 wireless vehicle detection installation by up to 1,000ft from the AP240-e.

A large database of three dimensional magnetic signatures of vehicles has been collected, with the help of these high quality sensor network products. They provide an invaluable source of data for detailed analysis. Below is a summary of the characteristics of the VSN240 family products.

Microcontroller	66MHz 5272 Coldfire processor
RF Specification	IEEE 802.15.4 standard compliant
	16 Channels in 2.4 - 2.48 GHz ISM band
Communication range	50m with AP240 at a height of 8m
	35m with AP240 at a height of 6m
	25m with AP240 at a height of 4m
	1000ft between RP240 and AP240
Expected lifetime	7 - 10 years (depending on application)
Operating temperature	-40C to 85C

Table 3.2.5.1 Summary of characteristics of the VSN240 family products

3.3 Communication Protocols and Lifetime Analysis

In a sensor node, more than 90% of the battery energy is consumed by the radio [3.10]. The network's communication protocol, which determines how the radios are operated, has a decisive influence on the system lifetime. In this section, the Sensys' Nanopower [3.18] and PEDAMACS [3.10] communication protocols are discussed in 3.3.1 and 3.3.2 respectively. The corresponding lifetime analysis is presented in 3.3.3.

3.3.1 Sensys' Nanopower Protocol

Existing Medium Access Control (MAC) protocols fall into one of two categories: random access schemes and Time Division Multiple Access (TDMA) schemes [3.25]. Random access schemes have the advantage of accommodating unpredictable need to transmit data. But a large portion of power is wasted by a node in idle listening, in overhearing packets that are not destined for it, and in collisions during packets transmission. Enhancements like SMAC [3.26] and PAMAS [3.27] were proposed to save power up to a factor of ten by reducing the radio active time, at a cost of considerable increase in hardware complexity.

The TDMA schemes are more power efficient on the other hand. The sensor nodes remain in an inactive state that consumes almost no power, until it is time for their allocated communication slot. However, general TDMA schemes do not take advantage of the fact that all data from sensor nodes is destined for a single access point (AP). So some communication power is wasted in distributing synchronization overhead [3.28].

The Sensys' Nanopower protocol [3.18] is a TDMA scheme designed to meet both the periodic communication and power saving requirements. It assumes that all the sensor nodes can reach either the AP or a repeater node in one hop, and all the repeater nodes can reach the AP in one hop. The AP, with its unlimited power supply, can communicate with all the sensor and repeater nodes in one hop. The simple communication link structure of this protocol is shown in Fig. 3.3.1.1.

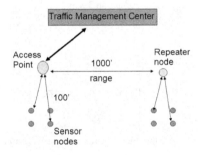

Fig. 3.3.1.1 Communication links structure of Sensys' Nanopower protocol

The communication protocol can be summarized as follows:

i, All the sensor and repeater nodes are configured with a pre-assigned radio channel and a communication time slot before installation.

ii, A synchronization message is send by the AP periodically with a random back off time, so all the reachable nodes can catch this message eventually.

iii, The radios of all the synchronized sensor nodes will wake up during its assigned time slot for two-way communication, and quickly switch back to sleep mode afterward.

iv, The repeater nodes are configured with the time slots of a multiple number of sensor nodes, so that they can wake up at the right time to collect data from these sensor nodes and transmit the data back to the AP at their own time slots.

v, After data processing is done at the AP, useful traffic data are transmitted to the Traffic Management Center (TMC) for further traffic controls.

The exclusion of multi-hop communication between sensor nodes greatly simplifies the protocol. The drawback is that the network coverage is limited by the maximum communication range between the AP and repeater nodes (e.g. 1000'). For the implementation of a traffic surveillance station, this coverage is more than enough in most cases. Using this Nanopower protocol with the VSN240 sensor network hardware family manufactured by Sensys Networks [3.18], it is claimed that a system lifetime of 10 years can be achieved for a typical vehicle detection application.

3.3.2 PEDAMACS

PEDAMACS (Power Efficient and Delay Aware Medium Access Protocol for Sensor Networks) [3.10] is a TDMA scheme that extends the single-hop structure to a multi-hop network. Unlike Sensys' Nanopower protocol [section 3.3.1] which requires the network configuration to be setup manually, the network topology is discovered by PEDAMACS automatically. The main assumption is that the AP has to be powerful enough to communicate with all the sensor nodes in one hop. However, multi-hop communication is used for data transmitted from sensor nodes to AP. A simple communication links structure of this protocol is shown in Fig. 3.3.2.1.

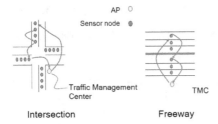

Fig. 3.3.2.1 Communication links structure of PEDAMACS

The protocol applies to a network in which sensor nodes periodically generate data to be transmitted to the AP. The protocol operates in four phases: topology learning, topology collection, scheduling, and adjustment:

i, In the topology learning phase, each node identifies its (local) topology, i.e. its neighbors, interferers, and its parent node in the routing tree rooted at the AP obtained according to some routing metric.

ii, In the topology collection phase, each node sends its local topology information to the AP so, at the end of this phase, the AP knows the full network topology.

iii, At the beginning of the scheduling phase, the AP broadcasts a schedule. Each node then follows the schedule, and sleeps during time slots when it is not scheduled to transmit a packet or to listen for one.

iv, The adjustment phase is triggered as necessary to learn the local topology information that was not discovered during the topology learning phase or to discover changes.

The determination of the schedule based on the topology of the network at the AP is performed according to the PEDAMACS scheduling algorithm [3.10]. The scheduling

56

algorithm should ideally minimize the delay—the time needed for data from all nodes to

reach the AP. However, this optimization problem is NP-complete. Instead, PEDAMACS

uses a polynomial-time scheduling algorithm which guarantees a delay proportional to the

number of packets in the sensor network to be transferred to the AP in each period. The

algorithm assigns a group of non-conflicting nodes to transmit in each time slot, in such a

way that the data packets generated at each node reaches the AP by the end of the

scheduling frame. A comparison of the delay in different communication protocols with

different number of sensor nodes is shown in Fig. 3.3.2.2 [3.10].

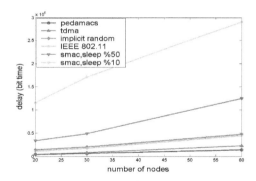

Fig. 3.3.2.2 Comparison of data transmission delay in different communication protocols
[3.10]

Although the protocol is designed for the situation in which every sensor node periodically

generates data to be transmitted to one AP, the PEDAMACS framework is flexible enough

to be generalized in many ways. It can be extended to networks with event-driven data

generation, multiple APs, as well as handling sensor nodes that are outside the range of a

single hop from the AP. This allows the protocol to be deployed in different applications of

traffic surveillance. The lifetime analysis of PEDAMACS is presented in the next section.

3.3.3 Lifetime Analysis

The lifetime analysis of different communication protocols was conducted by simulating a typical wireless sensor network using TOSSIM [3.29] –a sensor network simulation framework developed in the TinyOS project [3.20]. The following network configuration was adopted in the simulation: 50 kbps transmission rate, 128Hz sampling rate, 2 minute packet generation period, 2200 mAh@3V power supply from a pair of AA batteries. Power consumption of basic operations in MICA sensor nodes [3.30] shown in Table 3.3.3.1 was used for calculating the system lifetime. Six communication protocols shown in Table 3.3.3.2 were tested in the simulation.

Operation	Power Consumption
transmitting one packet	0.92mJ
receiving one packet	0.69mJ
listening to channel	29.71mJ/sec
operating radio in sleep mode	15μJ/sec
sampling sensor	1.5μJ/sample

Table 3.3.3.1 Power consumption of basic operations in MICA sensor node [3.30]

Index	Protocol
P1	PEDAMACS
P2	Implicit random
P3	IEEE 802.11, random access scheme with explicit acknowledgements
P4	SMAC [3.26] with 50% duty cycle
P5	SMAC [3.26] with 10% duty cycle

Table 3.3.3.2 Communication protocols tested in the TOSSIM simulations

The simulated system lifetimes for different number of sensor nodes are shown in Fig. 3.3.3.1. The lifetime of random access schemes (P2, P3) is about 10 days, whereas the one of P5 increases up to 60 days. On the other hand, PEDAMACS gives a lifetime of about 1200 days. The reason for the dramatic difference is clearly shown in Fig. 3.3.3.2, which compares the power consumed by these schemes in different operations for a 60-sensor node network (y-axis is presented in log scale).

The primary cause of the difference in lifetimes is in the total energy consumed by the radio in 'listening' and 'reception' operations. P5 can decrease this energy by a factor of 10 whereas PEDAMACS decreases it by a factor of more than 1,000. The difference in lifetimes also arises from the differences in amount of energy spent in transmission due to retransmissions and reception because of the 'overhearing effect': In random access schemes, when one node transmits a packet, all neighboring nodes receive it whereas only the parent of that node receives it in PEDAMACS.

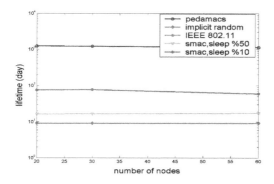

Fig. 3.3.3.1 Simulated system lifetime for different number of sensor nodes with different communication protocols [3.10]

59

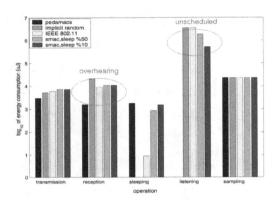

Fig. 3.3.3.2 Simulated power consumptions in different operations for a 60-sensor nodes network with different communication protocols [3.10]

In order to make such a sensor network competitive with the inductive loop detectors which have a lifetime of ~10 years [Ch.2], the lifetime of the PEDAMACS network can be further increased by using extra repeater nodes in conjunction with energy efficient routing that balances the energy consumption on multiple paths. In [3.31], this is formulated as a linear programming problem, by restricting the locations where the repeater nodes may be placed on a square lattice. Given the location of sensor nodes, the optimal locations of repeater nodes together with the optimal energy provided to them can be determined. The simulation results of a simple example are shown in Fig. 3.3.3.3. Using extra repeater nodes in the network allows the transmission range to be reduced from 40ft to 20ft, and making one unit of battery energy to be enough to achieve the target 10-year lifetime.

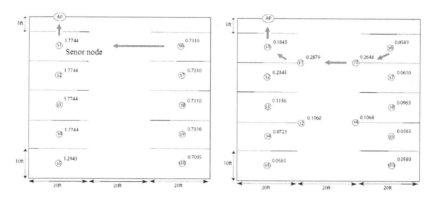

(a) Sensor nodes with 40ft transmission
range

(b) Using repeater nodes with 20ft
transmission range

Fig. 3.3.3.3 Simulated battery energy required for achieving 10-year system lifetime [3.31]

61

Ch. 4 Vehicle Detection by Wireless Sensor Networks

Traffic surveillance starts with the capability to detect the presence of a vehicle. Based on such detection, statistics like vehicle counts, traffic flow speed and occupancy are calculated. The quality of the vehicle detection determines the performance of all the dependent applications. In this chapter, signal analyses of two types of sensors, acoustic and magnetic, which can potentially be used in sensor nodes are discussed in section 4.1 and 4.2. An efficient and robust real-time detection algorithm for these sensor signals, called Adaptive Threshold Detection Algorithm [4.1] is studied in section 4.3. Finally, the experimental results and analysis of such a vehicle detection system are presented in section 4.4.

4.1 Acoustic Signal Analysis

At the early stage of this research project, the feasibility of using an acoustic sensor for vehicle detection was studied (jointly with Jiagen Ding, a graduate student supervised by Prof. Pravin Varaiya). In this section, the acoustic sensor used and corresponding signal processing are discussed.

4.1.1 Acoustic Sensor

Fig. 4.1.1.1 Pictures of MICA2, MICA2DOT and sensor board MTS310 (Left to Right)

The sensor node used for this analysis was MICA2 [4.2] with sensor board MTS310 [4.3].

MICA2 is another sensor node platform in the same Crossbow Mote family as

MICA2DOT [3.15] discussed in Ch. 3. Both MICA2 and MICA2DOT have the same

characteristic but in different form factors.

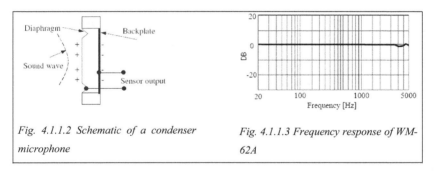

Fig. 4.1.1.2 Schematic of a condenser microphone

Fig. 4.1.1.3 Frequency response of WM-62A

The acoustic sensor on the sensor board MTS310 is a Panasonic WM-62A microphone

[4.5], which is an omni-directional condenser type microphone with frequency response

shown in Fig. 4.1.1.3. The schematic of a typical condenser microphone [4.6] is shown in

Fig.4.1.1.2. It includes a stretched metal diaphragm that forms one plate of a capacitor. A

metal disk placed close to the diaphragm acts as a back-plate. A stable DC voltage is

applied across the plates through a high resistance to keep electrical charges on the plates.

When a sound wave excites the diaphragm, the capacitance between the two plates varies according to the variation in the sound pressure. The change in the capacitance generates an output proportional to the sound pressure, which is at an ultra low-frequency pressure variation. A high-frequency voltage (carrier) is applied across the plates and the acoustic sensor output signal is the modulated carrier. Fig. 4.1.1.4 shows an example of a vehicle acoustic signal collected from this condenser microphone.

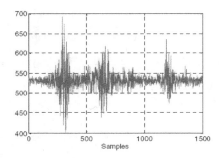

Fig. 4.1.1.4 Example of vehicle acoustic signal collected from the condenser microphone

4.1.2 Background Noise

To better understand vehicle acoustic signals in the frequency domain, experiments were conducted in the Richmond Field Station (RFS) with an omni-microphone (Radio Shack 33 − 3025A) connected to a laptop computer sampling at 11 kHz with 8-bit resolution. Short-Time Fast Fourier Transform (SFFT) [4.7] analysis was conducted on the acoustic data collected under different conditions. The N-point SFFT in the discrete domain is defined as

$$X(n, k) = \sum_{m} x(n)w(n - m)e^{-j2\pi mk/N}$$

(Eq. 4.1.2.1)

64

x(n) is the signal for analysis and w(n) is the window function. Popular window functions include rectangular and hamming windows. SFFT can be interpreted as a sequence of discrete time Fourier Transforms as the window w(n – m) slides along the signals [4.8]. In the following SFFT analysis, a hamming window is used.

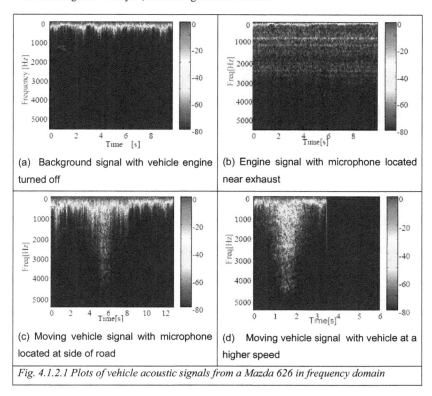

(a) Background signal with vehicle engine turned off	(b) Engine signal with microphone located near exhaust
(c) Moving vehicle signal with microphone located at side of road	(d) Moving vehicle signal with vehicle at a higher speed

Fig. 4.1.2.1 Plots of vehicle acoustic signals from a Mazda 626 in frequency domain

The results of SFFT analysis on the acoustic signals of a Mazda 626 are shown in Fig. 4.1.2.1 above. Fig.4.1.2.1(a) shows the background acoustic signal in frequency domain, with the vehicle remaining stationary and engine turned off. Similar experiments were conducted at a number of different test locations. It was observed that the magnitude of the background noise depends on the environment and level of wind. However, the background noise stays below 500 Hz. Fig.4.1.2.1(b)(c)(d) show the vehicle acoustic signal in the frequency domain under different situations. It was found that the acoustic signal of most passenger vehicles stayed below 5000Hz in the frequency domain.

4.1.3 Signal Processing

Raw data

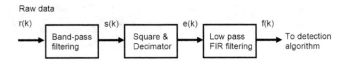

Fig. 4.1.3.1 Block diagram of the signal processing for the acoustic data

According to the SFFT results, a band-pass filter from 500Hz to 5000Hz was designed to filter out the background noise. Fig. 4.1.3.2 demonstrates how the band-pass filter successfully filtered out the background and kept the noise at a small level. The band-pass filter output s(k) is squared and decimated, such that it is transformed into an energy distribution signal e(k). It is related to s(k) by:

$$e(k) = [s(Nk)]^2 \qquad \text{where N is the decimating rate} \qquad \text{(Eq. 4.1.3.1)}$$

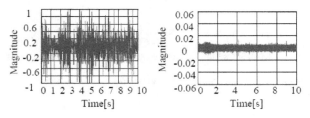

Fig. 4.1.3.2 Example of background acoustic in time domain before (left) and after (right)
band-pass filtering [4.1]

Since the energy signal e(k) can still be jerky, a smoothing filter is introduced. A low-pass
FIR filter [4.9] is used for the smoothing because of its linear phase and inherent stability.
The key design parameters for this low-pass FIR filters are the -3dB cutoff frequency (ω_p),
the stop band frequency (ω_s) and the stop band attenuation gain. Fig. 4.1.3.3 shows the
magnitude response of the FIR filter being used. At the end, the smoothed energy signal
f(k) is passed to the detection algorithm [section 4.3] for detection decision making. Fig.
4.1.3.4 shows a sample vehicle acoustic signal from a traveling Mazda 626 at different
processing steps.

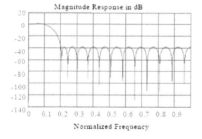

Fig. 4.1.3.3 Magnitude response of the FIR filter

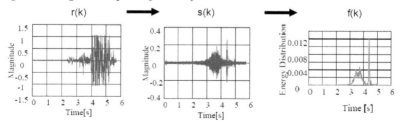

Fig. 4.1.3.4 Vehicle acoustic signal from a traveling Mazda 626 at different processing
steps [4.1]

67

4.2 Magnetic Signal Analysis

4.2.1 Simulation

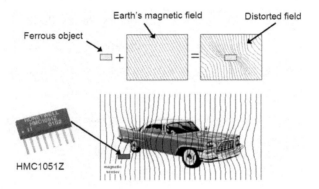

Fig. 4.2.1.1 Basic concept of using a magnetic sensor for detecting vehicle

As noted in Ch.3, the AMR magnetic sensor [3.14] detects the distortions of the Earth's field which is uniform over a wide area on the scale of kilometers. Depending on the composition of the ferrous material inside a vehicle, its size and orientation, different magnetic signatures are induced by vehicles with different configurations and shapes. Such distortions caused by a large ferrous object like a vehicle can be modeled as a composite of many dipole magnets [3.19].

Maxwell's equations are a set of fundamental equations governing electromagnetism. A reduced form of the Maxwell's equations, the magnetostatic equations (Eq. 4.2.1.1), are sufficient to describe the interaction in our simulation.

$$\nabla \cdot B = 0$$
$$\nabla \times B = \mu_0 \mu_r J$$

(Eq. 4.2.1.1)

B is the magnetic flux density, J is the current density, μ_0 is the permeability of air and μ_r is the relative permeability of the medium. The magnetostatic equations are valid only if all the electric charge densities are constant and the currents are steady, so that the electric and magnetic fields are static. Since the magnetic field is changing slowly when a vehicle passes, it can be simulated by the magnetostatic equations.

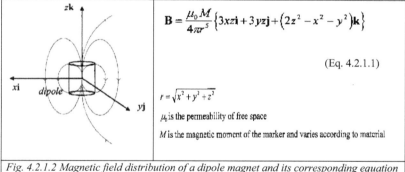

$$\mathbf{B} = \frac{\mu_0 M}{4\pi r^5}\left\{3xz\mathbf{i} + 3yz\mathbf{j} + \left(2z^2 - x^2 - y^2\right)\mathbf{k}\right\}$$

(Eq. 4.2.1.1)

$r = \sqrt{x^2 + y^2 + z^2}$

μ_0 is the permeability of free space

M is the magnetic moment of the marker and varies according to material

Fig. 4.2.1.2 Magnetic field distribution of a dipole magnet and its corresponding equation

Fig. 4.2.1.2 shows the magnetic field strength (B) distribution in 3D space of an ideal dipole magnet, given by the equation on its right. A simulation of the magnetic vehicle signature was conducted according to the experimental configuration shown in Fig. 4.2.1.3. The vehicle was simulated as a plate array of point-source dipole magnets. The simulated magnetic signature of such a vehicle with a 3 –axes magnetic sensor located at the side of lane, traveling from x- to x+ direction as shown in Fig. 4.2.1.4(a). On the other hand, experimental results conducted under similar configuration can be found in Fig. 4.2.1.5(b).

A Ford Taurus was used as the test vehicle, with magnetic sensor sampling at 64Hz for each of the axis. Even though the magnitude levels and units between the two plots are different because the simulation did not account for the magnetic properties of the test vehicle, the simulated patterns are in agreement with the experimental measurements.

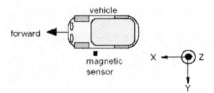

Fig. 4.2.1.3 Experimental configuration for the magnetic vehicle signature simulation

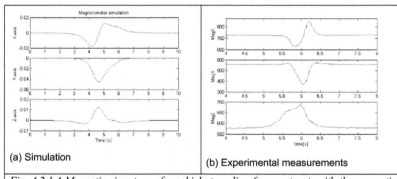

(a) Simulation

(b) Experimental measurements

Fig. 4.2.1.4 Magnetic signature of a vehicle traveling from x- to x+ with the magnetic sensor located at the side of lane

4.2.2 Drifting of Magnetic Measurements

It is known that the measurements from HMC1051Z magnetic sensor [3.14] are affected by temperature. This could be an issue for traffic surveillance as the temperature on the pavement can change a lot in the course of a day. Therefore, an experiment was carried out to study the effect of direct sunlight on the sensor node's magnetic measurements.

Fig. 4.2.2.1 Experimental setup for studying the effect of direct sun light on magnetic measurements

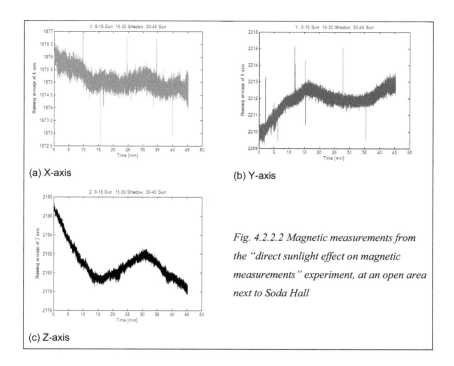

(a) X-axis

(b) Y-axis

Fig. 4.2.2.2 Magnetic measurements from the "direct sunlight effect on magnetic measurements" experiment, at an open area next to Soda Hall

(c) Z-axis

71

The experimental setup (Fig. 4.2.2.1) was simple: a sensor node was placed in an open area next to Soda Hall where there is sufficient direct sun light. This experiment was conducted at noon in March 08, 2005, with temperature at ~60F. A total of 45 minutes of 3-axes raw magnetic measurements were collected. The sensor node was under direct sunlight for the first 15 minutes. A shadow over the sensor node was created manually with an umbrella 2m above the ground for the next 15 minutes. The shadow was removed and the sensor node was under direct sunlight again for the last 15 minutes. The X, Y and Z-axis measurements are shown in Fig. 4.2.2.2.

The Z-axis measurements show a very clear linear drift. The magnetic reading decreased from 1287 to 1279 during the first 15 minutes under sun light, then went back up to 2182 in the 15 minutes of shade and decreased again to 1286 in the last 15 minutes of sunlight exposure. The Y-axis measurements show a similar behavior but in the opposite direction. The X-axis measurements remained pretty much the same during the experiment. This up-and-down drifting behavior of the Z and Y-axis measurements clearly demonstrated the effect of sun light (temperature) on the magnetic sensor.

Another day-long experiment was carried out to study this drift on a day long basis. A sensor node was placed on the ground of the backyard of my apartment (Derby Street, Berkeley, CA 94705), where directly sunlight is not reachable most of the time. 24 hours of measurements were recorded starting from 5pm of March 05, 2005. Plots of measurements are presented in Fig. 4.2.2.3. The results agree with the "direct sunlight effect" experiment with the Z-axis reading going up slowly as the temperature decreased from 5pm to 7am., and vice versa during 7am to 2pm. There was a relative sharp charge in all the 3-axes measurements at 3pm, when there is a sudden exposure of direct sunlight. To summarize, the effect of temperature change on magnetic measurement, although unavoidable, it should be accounted for in the design of the vehicle detection algorithm.

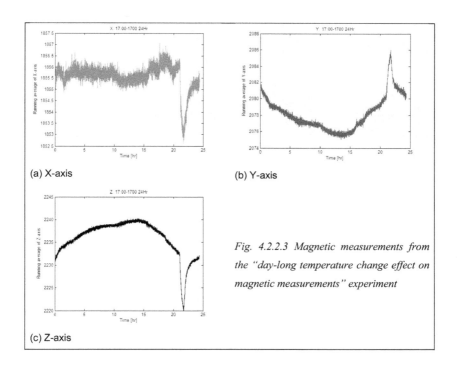

(a) X-axis

(b) Y-axis

(c) Z-axis

Fig. 4.2.2.3 Magnetic measurements from the "day-long temperature change effect on magnetic measurements" experiment

4.2.3 Signal Processing

The magnetic sensor is reliable. It has 1 bit of sensor noise out of a 10-bit signal resolution. So no extra signal processing step is necessary to reduce the noise, unlike in the case of acoustic signals. Since the frequent up-and-down fluctuation of the magnetic signal is not a desirable characteristic for the vehicle detection algorithm, a smoothing filter, which takes a running average of the signal, is used to smooth out the signal. The running average is given by

$$a(k) = \begin{cases} \dfrac{r(k) + r(k-1) + \ldots r(1)}{k} & \text{for } k < M \\[2mm] \dfrac{r(k) + r(k-1) + \ldots r(k-M+1)}{M} & \text{for } k >= M \end{cases}$$

(Eq. 4.2.3.1)

$r(k)$ is the raw signal and M is the pre-defined running average buffer size. The plots of the raw magnetic signal and the output of the smoothing filter for the same vehicle sample are shown in Fig. 4.2.3.1.

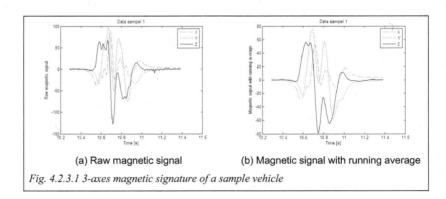

(a) Raw magnetic signal   (b) Magnetic signal with running average

Fig. 4.2.3.1 3-axes magnetic signature of a sample vehicle

4.3 Vehicle Detection Algorithm

Since this traffic surveillance system envisions a large scale deployment, the vehicle detection algorithm has to be sufficiently robust to provide accurate detection in different working environments. Moreover, it has to be computationally simple to be implementable with the limited capability of the sensor node's processor. With these two objectives in mind, an ad-hoc vehicle detection algorithm named Adaptive Threshold Detection Algorithm (ATDA) was designed for detecting vehicles in moving traffic, based on the magnetic signal. This algorithm turns out to be usable for the acoustic signal as well. Moreover, a modified version of ATDA can be used for detecting stationary vehicles in a parking lot. The details of these algorithms are discussed next.

4.3.1 Adaptive Threshold Detection Algorithm (ATDA)

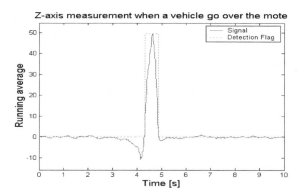

Fig. 4.3.1.1 A very simple vehicle signature of the Z-axis magnetic signal

The main reason for using a threshold detection approach instead of other statistical algorithms is to reduce the computational requirement of the detection algorithm so that it can be implemented on the sensor node's processor and generate detection results in real time. This could be easily done with a fixed threshold detection algorithm if all the vehicle signatures were as simple as the one shown in Fig.4.3.1.1. Unfortunately, the vehicle signature patterns can be much more complicated than that shown in Fig.4.2.3.1, which together with the drift discussed in section 4.2.2, make it difficult to design a simple detection algorithm. These considerations led to the Adaptive Threshold Detection Algorithm (ATDA). Whose block diagram is shown in Fig. 4.3.1.2. The Detection State Machine is shown in Fig. 4.1.3.3.

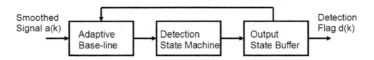

Fig. 4.3.1.2 Block diagram of the Adaptive Threshold Detection Algorithm

Adaptive Base-line

Even through there is an uncontrollable drift in the magnetic signal, the rate of the drift is quite small—on the order 1 measuring unit per minute, depending on the sensor node prototype being used. This implies that the drift has a negligible effect on the detection of a moving vehicle whose signature is about 1 second in duration. In order to account for the drift in the long term, an adaptive baseline is setup to track the background magnetic reading, which is used to determine the adaptive threshold level for the detection state machine. The adaptive baseline for each of the three magnetic axes is given by the following equations:

$$B_i(k) = \begin{cases} B_i(k-1) \times (1-\alpha_i) + a_i(k) \times (\alpha_i) & if\ s(\tau) = 0\ \forall\ \tau \in [(k - s_{buf})\ ...\ (k-1)] \\ B_i(k-1) & Otherwise \end{cases} \quad for\ i \in [x\ y\ z]$$

(Eq. 4.3.1.1)

$$T(k) = \begin{cases} \begin{cases} true & if\ |a_z(k) - B_z(k)| > h_z(k) \\ false & Otherwise \end{cases} & for\ s(k-1) \neq Event\_Detected \\ \begin{cases} true & if\ |a_z(k) - B_z(k)| > h_z(k)\ or\ |a_x(k) - B_x(k)| > h_x(k) \\ false & Otherwise \end{cases} & for\ s(k-1) = Event\_Detected \end{cases}$$

(Eq. 4.3.1.2)

B(k) is the adaptive baseline, α is the forgetting factor, a(k) is the smoothed magnetic data, s(k) is the state of the detection state machine, s_{buf} is the buffer size of s(k) and the subscript i represents one of the three axes. With this equation, the adaptive baseline is only updated by the magnetic reading when there is no signal fluctuation and no vehicle is detected within a certain period of time. The updating is slow with a forgetting factor of 0.05 to ensure that only the background magnetic signal is accounted for the adaptive baseline.

With this adaptive baseline, an "Over Threshold" Boolean flag is generated according to Eq. 4.3.1.2, and h(k) is the corresponding threshold level. From analysis of the empirical magnetic signal of an urban traffic stream, we know that there is a chance that a stop-and-go vehicle stays on top of a sensor node with its Z-axis measurements *below* the threshold level. In order to avoid double-counting such a vehicle, the x-axis measurements are also introduced into the decision of T(k) when the state machine is in its "Event_Detected" state. This effectively filters out the double-counting error as it is very unlikely that both the Z and X-axis measurements are below the threshold when a vehicle is present. Z-axis

77

magnetic measurements are used as the major source for vehicle detection because of its

localized characteristic, so that it can isolate detections from vehicle in adjacent lanes.

Detection State Machine

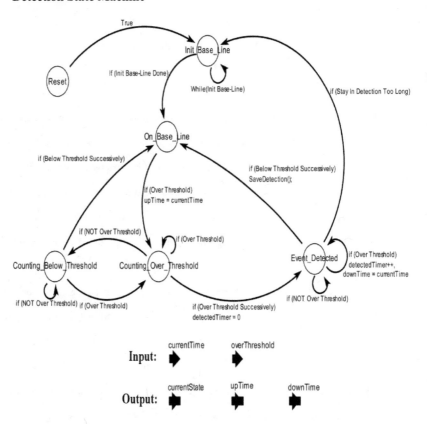

Fig. 4.1.3.3 State diagram of the ad-hoc state machine designed for ATDA

The "Over Threshold" Boolean flag is passed to an ad-hoc state machine as shown in Fig. 4.1.3.3 above. Its main objective is to filter out spurious signals that are not caused by a vehicle and to output binary detection flag without calling any complicated statistical function. The following section is a walkthrough of this state machine's logical flow.

S1   "Init_Base_Line"

Assuming there is no vehicle near the sensor node when it is being reset. It will go into state S1 and start initializing the baseline with the environmental measurements.

S2   "On_Base_Line"

After a pre-defined initializing time (namely, 3s), it will jump to the state S2 where the baseline is updated adaptively. It will jump to state S3 when a Z-axis measurement larger than the adaptive threshold is recorded. The potential up-time of this event is saved for later use in case this is a valid vehicle detection event.

S3   "Counting_Over_Threshold"

It was found that a vehicle signature produces a successive sequence of "True" in $T(k)$ and this state is used to track such a sequence. If there is any "False" reading from $T(k)$, it will immediately jump to state S4. Otherwise, if the number of successive $T(k)$ = "True" has reached a critical value, it will jump to state S5.

S4   "Counting_Below_Threshold"

Within this state, it will jump back to state S2 after the number of successive $T(k)$ = "False" has reached a critical value. In order not to lose a potential vehicle detection, it will jump back to the state S3 again in case there is any $T(k)$ = "True" reading.

S5   "Event_Detected"

Staying in this state implies the magnetic fluctuation is strong as the vehicle is still traveling over the sensor node. As discussed earlier in the generation of T(k), it is now determined by both the Z and X-axis data in this state. The down-time of this detection event will be updated up to the last T(k) = "True" reading. A complete vehicle detection event ends if the number of successive T(k) = "False" has reached a critical value and it will jump back to S2 and wait for another detection cycle. Moreover, in case there is any error in setting up the baseline and threshold causing the state machine to stay in this detected state for an unreasonably long period of time, a failsafe mechanism was introduced. Such a situation is identified by the detected timer when it exceeds a pre-defined value of longest reasonable detected time. It will jump back to S1 and reset the whole state machine.

Finally, the detection flag d(k) is generated according to the output state of this state machine. At a finer scale, the change in detection flag occurs within 10 measurements, i.e. in less than 0.1s immediately after the magnetic frontier of the vehicle crosses the sensor. Thus the detection can be reported within 0.1s to the controller which is a requirement for a traffic detection system at an intersection [4.10]. This computationally simple detection algorithm can be executed in the senor node. The detector cards used with the loop detector, which perform a similar vehicle detection function, are no longer needed. The performance of this ATDT is discussed in a later section of this chapter.

4.3.2 Speed and Magnetic Length Estimation

Speed Estimation by a Sensor Node Pair

The common practice for speed estimation using a single inductive loop detector is to assume a fixed, pre-defined magnetic vehicle length (e.g. 5 m) and calculate the speed from $V_{fixedLenght} = L_{fixed} / T_{occupancy}$ (Eq. 4.3.2.1). So the accuracy of the speed estimate depends on the difference between the actual vehicle length and the pre-defined one. More accurate estimation can be obtained using dual inductive loops or other road side sensors like radar and microwave. With a pair of sensor nodes, speed can be estimated in a manner similar to dual inductive loops.

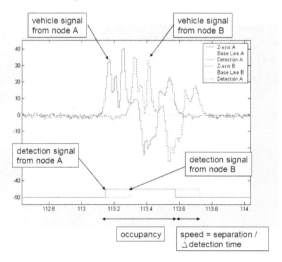

Fig. 4.3.2.1 Example of speed estimation by a pair of sensor node

Fig. 4.3.2.1 shows an example of speed estimation of one vehicle by a pair of synchronized sensor nodes. ("Synchronized" nodes mean that their clocks are synchronized.) Node A and

B were placed on the middle of a lane with a known separation of 6 feet along the traveling direction. Assuming the vehicle has negligible lateral offset and acceleration within these 6 feet, the vehicle signature measured by node A should be identical to the one measured by node B. Moreover, the detection flag should change virtually at the same point within the vehicle signature. As a result, the time difference between A and B at which the detection flags change is the travel time across the separation distance. The vehicle speed and their magnetic lengths are estimated by the following equations:

$$\Delta t_{up} = t_{B,up} - t_{A,up} \qquad \Delta t_{down} = t_{B,down} - t_{A,down}$$

$$v_{up} = \frac{D_{AB}}{\Delta t_{up}} \qquad\qquad v_{down} = \frac{D_{AB}}{\Delta t_{down}}$$

$$\hat{v}_{pair} = avg\left(v_{up}, v_{down}\right)$$

$$t_{A,occupancy} = t_{A,down} - t_{A,up} \qquad t_{B,occupancy} = t_{B,down} - t_{B,up}$$

$$t_{pair,occupancy} = avg\left(t_{A,occupancy}, t_{B,occupancy}\right)$$

$$\hat{v}_{fixedLength} = \frac{L_{fixed}}{t_{pair,occupancy}}$$

$$\hat{L}_{pair} = \hat{v}_{pair} \times t_{pair,occupancy} \qquad\qquad\qquad \text{(Eq. 4.3.2.2)}$$

t is time, v is speed, D is distance and L is magnetic length. The 'up' and 'down' subscripts refer to the change in the detection flag from 0 to 1 and from 1 to 0 respectively. Please be reminded that the magnetic length is usually longer than its physical length depending on the metallic composition of the vehicle.

If there is a small difference in sensitivity between nodes A and B, a difference in the amplitude of the vehicle signatures is induced. This will violate the presumption that the detection flag will change virtually at the same point within the vehicle signature. The node with a higher sensitivity will have a longer $t_{occupancy}$. Assuming that this sensitivity difference affects the timing of Δt_{up} and Δt_{down} to the same level but in opposite direction, its effect can be eliminated theoretically according to the following equations, in which $\overline{\Delta t}_{up}$ and $\overline{\Delta t}_{down}$ are the corresponding changes in time with the sensitivity effect ε.

$$\overline{\Delta t}_{up} = \Delta t_{up} - \varepsilon \qquad \overline{\Delta t}_{down} = \Delta t_{down} + \varepsilon$$

$$\overline{v}_{up} + \overline{v}_{down} = \frac{D_{AB}}{\Delta t_{up} - \varepsilon} + \frac{D_{AB}}{\Delta t_{down} + \varepsilon} = \frac{D_{AB}\left(\Delta t_{down} + \varepsilon + \Delta t_{up} - \varepsilon\right)}{\Delta t_{up}\Delta t_{down} + \varepsilon\left(\Delta t_{up} - \Delta t_{down}\right) + \varepsilon^2}$$

$$\simeq \frac{D_{AB}\left(\Delta t_{down} + \Delta t_{up}\right)}{\Delta t_{up}\Delta t_{down}} = v_{up} + v_{down}$$

where $\qquad \varepsilon^2 \simeq 0 \qquad and \qquad \left(\Delta t_{up} - \Delta t_{down}\right) \simeq 0$

(Eq. 4.3.2.3)

The major remaining errors are due to the synchronization error between the node pair clocks and the estimation errors caused by the finite sampling rate and the node separation. In fact, there is a tradeoff between accuracy of the speed estimate and the power consumption needed for more accurate synchronization and higher sampling rate. Please refer to section 3.3 for a discussion of time synchronization of the wireless sensor networks.

Median Speed Estimation by a Single Sensor Node

If the speed of each individual vehicle is not required for a specific traffic control application, a statistical estimate of the traffic flow speed can be obtained by a *single* sensor node. Following the argument of Coifman et al [4.11], assume that vehicles in a platoon have a similar speed $V_{platoon}$. If there are N vehicles in this platoon, the (N+1) unknowns $L_1, \ldots L_N$ and $V_{platoon}$ should satisfy N equations:

$$t_{occupancy,i} = L_i \times V_{platoon} \quad \forall i \in [1 \ldots N] \qquad \text{(Eq. 4.3.2.4)}$$

As shown in [4.11], if the probability distribution of the magnetic length, p(L), and $t_{occupancy}$ are known, a maximum likelihood estimate of $V_{platoon}$ can be calculated. A robust, alternative estimate of $V_{platoon}$ and L is given by

$$\overline{V}_{platoon} = \frac{\overline{L}}{\overline{t}_{occupancy}} \qquad \hat{L}_i = \frac{t_{occupancy,i}}{\overline{V}_{platoon}} \quad \forall i \in [1 \ldots N] \qquad \text{(Eq.4.3.2.5)}$$

$\overline{L}$ and $\overline{t}_{occupancy}$ are the median magnetic length and median occupancy time respectively. Statistically, it was found that the standard deviation σ of a 10-point (i.e. N = 10) median speed estimate is 2.5 mph. This implies that with a probability of 0.95, the speed estimate error is less than 2 σ or 5 mph. The experimental results of such an analysis are discussed in the next section.

4.4 Experimental Results and Analysis

Several experiments focusing on vehicle detection, speed and magnetic length estimation are discussed in this section: in section 4.4.1 (dataset D1), preliminary results of vehicle detection by acoustic sensors are shown; in section 4.4.2 (dataset D2), the extraction of different traffic information from a single sensor node is studied; in section 4.4.3 (dataset D3), the experimental results of speed and magnetic length estimation from a sensor node pair are discussed; in section 4.4.4 (dataset D4), the experimental results of speed estimation are compared with video-based estimation; in section 4.4.5 (dataset D5), field test data are used to compare inductive loop detectors and magnetic sensor nodes.

4.4.1 Experiments with Acoustic Sensors (Dataset D1)

This section presents vehicle detection results based on the acoustic signal, and with the signal processing and ATDA implemented on a laptop computer. The acoustic signal is sampled at 11 kHz. The system was not implemented on the sensor node because it lacked the computation capability needed for the signal processing discussed in section 4.1.3. However, an FPGA chip can be used to implement the signal processing design and work with directly with the sensor node [4.12].

Fig. 4.4.1.1 shows the results of a real time vehicle detection experiment by acoustic signal, conducted on the side of a lane in local traffic. 7 out of 7 vehicles were detected successfully with the ground truth recorded manually. In order to test the robustness of the detection algorithm with a limited acoustic database, the data set collected from Richmond Field Station, Berkeley Marina Parking Lot and a local intersection on Hearst were mixed in random order to create a 'new' data set for testing. The corresponding results are summarized in Table 4.4.1.1, and give an error in vehicle counts below 7% in all off-line tests.

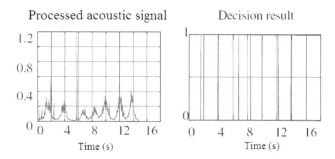

Fig. 4.4.1.1 Results of a real time vehicle detection experiment by acoustic signal

Case number	Ground truth (# of vehicles)	Detection result (# of vehicles)	Error in Vehicle Counts %
1	63	60	4.76
2	63	61	3.17
3	63	64	1.59
4	63	67	6.35
5	63	64	1.59

Table 4.4.1.1 Results of off-line testing with randomly mixed acoustic signal

Although these preliminary results are based on a very small data pool, they do suggest that acoustic sensors data could be processed so as to meet a vehicle count error rate under 5 percent, which would be necessary to be considered as a viable alternative to loops.

4.4.2 A Single Sensor Node Experiment in Local Traffic (Dataset D2)

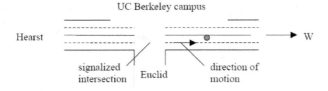

Fig. 4.4.2.0 Layout of the experimental site of the single magnetic sensor node experiment at Hearst Avenue (dataset D2)

One sensor node sampling at 128Hz was placed in the middle of a lane on Hearst Avenue, Berkeley, CA, at February 23, 2004, 8-9 pm. The setup was located on a section right after a traffic light-controlled intersection. Ground truth was established by a visual count. A total of 332 vehicles were observed, with the following distribution:

Type	Passenger	SUV	Van	Pickup	Bus	Motorcycle
Counts	248	48	18	9	4	5

Table 4.4.2.1 Distribution of vehicle types

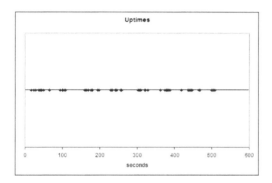

Fig. 4.4.2.1 Arrival time of vehicles during first 10 minutes in dataset D2

Detections were generated at real time by the magnetic sensor node without any post-processing. A total of 330 out of 332 (99%) vehicles were detected. The two undetected vehicles were motorcycles, so all non-motorcycle vehicles are detected. Three motorcycles that passed near the node were detected, so placing more nodes laterally will ensure motorcycle detection. In fact, more information can be extracted because each vehicle is individually measured. Fig. 4.4.2.1 shows a plot of vehicle arrival time during the first 10 minutes. The arrivals are bunched together into 'platoons' formed by the clearing of a queue behind the traffic signal during each green phase. Successive platoons are one

minute apart, which agrees with the cycle time of the traffic signal. The variability in

platoon size implies that the queue is cleared during each green phase and so one may

deduce that the traffic is not saturated. Typical signal control detection systems do not

measure traffic downstream of a signal. But this figure shows that such measurements can

reveal how well the signal plan is adapted to the traffic demand.

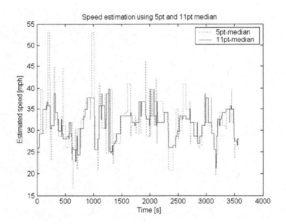

Fig. 4.4.2.2 Vehicle speed estimated by the 5-point and 11-point median approach in
dataset D2

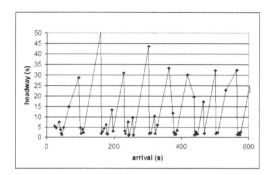

Fig. 4.4.2.3 Headway against arrival time of vehicle in dataset D2

As discussed in section 4.3.2, speed can be estimated with a single sensor node using the median speed approach. Taking the parameter values suggested in [4.11], $\overline{L}$ = 5m, N = 11 was used and N = 5 was chosen for comparison. Fig. 4.4.2.2 displays the corresponding results. The 11-point estimate is smoother than the 5-point estimate, as expected. In the experiment, the traffic flow was 330 vehicle/hr. Hence the passage of 11 vehicles takes about 2 minutes, so the 11-point estimate corresponds to a 2-min average. Under a heavier traffic flow, say 2,000 vehicle/hr, this will be a 20-second average.

The data also give the inter-vehicle headway by subtracting the up-time (arrival) of a vehicle from the downtime (departure) of the preceding vehicle. Fig. 4.4.2.3 plots the headway in seconds for the first 10 minutes. Grouped headway with large separation is clearly observed, which is caused by the departure of platoons created by the traffic signal. Applying Eq.4.3.2.5, gives an estimate of magnetic lengths whose empirical distribution is shown in Fig. 4.4.2.4.

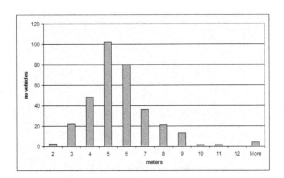

Fig. 4.4.2.4 Distribution of estimated magnetic length in dataset D2

4.4.3 A Sensor Node Pair Experiment in Local Traffic (Dataset D3)

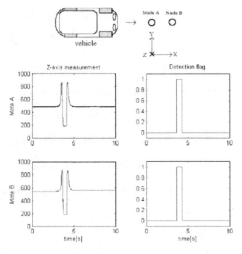

Fig.4.4.3.1 Z-axis measurements of a vehicle running over nodes at 16 mph

Fig. 4.4.3.1 shows the setup for a simple node pair experiment with nodes separated by 6.5 [inches]. With a sampling rate of 256 Hz, the speed estimated with the node pair method is 15.76 mph in this example, in agreement with the reference speed of 16mph from the GPS speed measurement on the vehicle. A real time traffic experiment was conducted with a similar configuration as in the example above. A one-hour long data trace was collected at a location similar to that in the previous Hearst Avenue experiment, with a sampling rate of 128 Hz and a node separation of 1.5 m. A total of 333 vehicles were counted manually during the test period. A summary of the experimental results is shown in Table 4.4.3.1. Over 97% of vehicles are detected by both nodes. Fig. 4.4.3.2 shows the distribution of estimated speed by the conventional fixed effective length approach.

	#Detections	Mean	STD
Manual Counting	333	—	—
Speed estimated by node A	330(99%)	30.0mph	10.4mph
Speed estimated by node B	326(97.9%)	33.8mph	13.6mph
Speed estimated by node pair	238(71%)	30.9mph	12.8mph
Length estimated by node pair	238(71%)	5.1m	2.5m

Table 4.4.3.1 Summary of experiment result in dataset D3

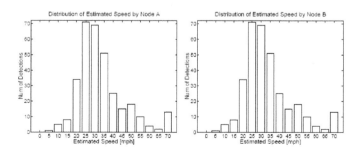

91

Fig. 4.4.3.2 Distribution of estimated speed by the conventional fixed effective length approach in dataset D3

For estimating speed with the node pair approach, those samples with unaligned "up" and "down" time pairs between nodes A and B were dropped. Samples are sometimes unaligned because of the error caused by synchronization and sampling resolution (These problem are fixed in the new Sensys sensor nodes [3.18]). A total of 238 (71%) valid detection pairs are identified. The resulting estimated speed and magnetic vehicle length distributions are shown in Fig. 4.4.3.3. Fig. 4.4.3.4 compares the distributions of estimated speed obtained from the fixed effective length approach and the node pair approach. The estimates do agree with each other and a reasonable distribution is observed except for a small number of exceptional cases with extremely high speed and short estimated magnetic length. Fig. 4.4.3.5 can be used to compare the distributions of the estimated vehicle magnetic length and of vehicle types observed during the test. Although we do not have the ground truth for the vehicle length in this test, the estimated distribution of vehicle length does agree with the vehicle type distribution.

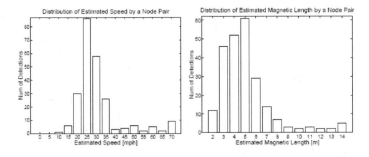

Fig. 4.4.3.3 Distribution of estimated speed and magnetic vehicle length by the node pair approach in dataset D3

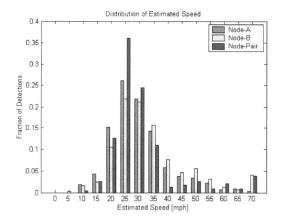

Fig. 4.4.3.4 Comparison between the distribution of estimated speed from the fixed effective length approach and node pair approach in dataset D3

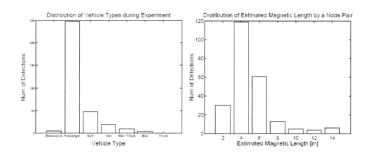

Fig. 4.4.3.5 Comparison between the distribution of the estimated vehicle magnetic length and that of vehicle types observed in dataset D3

4.4.4 Speed Estimation Comparison with Video (dataset D4)

Fig. 4.4.4.1 Picture of the experimental setup for dataset D4

In this experiment, two sensor nodes were placed six feet apart on Hearst Avenue. Two cones were placed 22.6' apart on the site to be used as landmarks for video processing. The speeds estimated by the sensor node pair and by the video were compared.

Statistic	Video (mph)	SN (mph)
Average	29.2	28.8
Minimum	20.1	19.1
Maximum	46.3	46.0
Median	28.9	28.5

Table 4.4.4.1 Comparison of estimated speeds from a sensor node pair and video in dataset D4

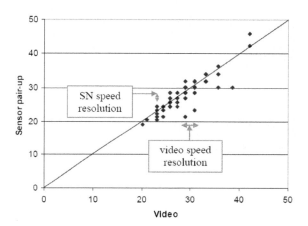

Fig. 4.4.4.2 Comparison of speeds determined by two sensor nodes and the video in dataset D4

Summary statistics of the speeds estimated by the two methods are shown in Table 4.4.4.1 above. The scatter plot of the two speed estimates is shown in Fig. 4.4.4.2. The overall statistics of the estimates from the two methods agree with each other. Since the sampling rate of the magnetometer is 128Hz while the video frame rate is only 30Hz, the video estimates are less accurate and have a positive bias, even though the landmarks for the video are much further apart than the sensor nodes.

4.4.5 Vehicle Detection Comparison with Inductive Loop

Detectors (Dataset D5)

Magnetic Sensor VS Inductive Loop

As discussed in section 3.2.1, the magnetic sensor HMC1051Z [3.14] is a passive device that measures the strength and direction of the Earth's magnetic field. By contrast, the inductive loop is an active device: a 6' by 6' copper loop is excited by a 20 kHz voltage in order to creating a magnetic field [4.13]. Conducting material passing over the loop lowers the inductance and the change in inductance is measured by an electrical detector card used with the loops. Special high scan-rate detector cards used for vehicle classification sample the inductance at 140Hz.

Another difference stems from the fact that the magnetic sensor node measures a highly localized change. As the vehicle travels over the sensor, it records the changes in the fields caused by different parts of the vehicle. By contrast, the 6' by 6' standard loop geometry results in the "integration of the inductive signature over the traversal distance", which can remove distinctive features from the inductive signature [4.14]. So the standard loop is not ideal for vehicle classification. Fig. 4.4.5.1 reproduces the inductive loop signatures of a pickup truck and a passenger car. A comparison with Fig. 4.3.2.1 clearly shows that the magnetic sensor node measurement provide much more detail than an inductive loop signature.

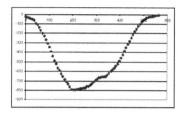 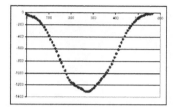

Fig. 4.4.5.1 Inductive loop signature from a pickup truck (left) and a passenger car (right): Source [4.15]

Detection Performance Comparison

Fig. 4.4.5.2 Picture of the experimental setup for dataset D5

An experiment was conducted at Oct 6, 2004 on a local traffic lane of Martin Luther King Way (MLK) in downtown Berkeley from 1:20pm to 3:20pm (119min). A Sensys sensor node [3.18] was placed on the pavement in the middle of an inductive loop located at a section before a traffic light-controlled T-intersection. With help from the City of Berkeley, real time data were collected from both detection systems for detailed analysis. Moreover, video of the traffic was captured to be used as the source for ground truth.

97

	# Detections	Correct counts	[%]
Video	791		100
Inductive loop	904		114.2857
Sensor node	791+7-7 = 791	791-14 = 777	98.23009
over-counting	7		
Adj. lane		4	
Double counted		2	
Packet loss		1	
under-counting	7		
Level of measurements not high enough		3	
Changing lane / Not along middle of lane		2	
Motorcycle		1	
Packet loss		1	

Table 4.4.5.1 Summary of vehicle detection results of dataset D5

The vehicle detection results are summarized in Table 4.4.5.1. A total of 791 vehicles were observed in the recorded video. 904 detection events were generated by the inductive loop detector, which over-counted by 14%. This can be explained by the fact that this loop detector was used as a presence detector instead of a traffic counter, so it may not be well calibrated for counting vehicles. Since the traffic is quite heavy at MLK, a large number of stop-and-go cases were observed. This could be the main source of error for double counting by the loop detector.

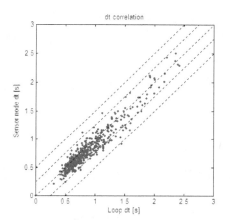

Fig. 4.4.5.3 Correlation of occupancy time for each individual detection between the sensor node and loop detector in dataset D5

On the other hand, the sensor node shows a virtually 100% correct detection of the overall vehicle counts. But if we examine the detection of individual vehicles, 7 over-counting and 7 under-counting instances were identified. The causes of these errors are also summarized in the Table 4.4.5.1. Even after subtracting these 14 cases, a very high successful detection rate of 98.2% was achieved. A plot of the correlation of occupancy time between the sensor node and loop detector is shown in Fig. 4.4.5.3. An overall correlation coefficient of 0.67 was obtained. This is caused by the difference in zone of detection as discussed at the beginning of this section. With this promising result from an urban traffic intersection with heavy traffic flow, one may confidently predict that such a robust wireless sensor network can detect vehicles and estimate speeds as well as a highly calibrated inductive loop detector.

Ch. 5 Vehicle Classification by Wireless Sensor Networks

Vehicle classification refers to the process and methodology to classify a vehicle signature in a specific format into a pre-defined vehicle class (e.g. passenger vehicle or truck). It is an important source of information for transportation design and management that can be used for many purposes. In pavement design and management, pavement life is estimated according to the distribution of vehicle types running over it, and this distribution may be used to schedule re-surfacing. In traffic safety research and implementation, the distribution of trucks in traffic is a critical design factor, because of its significantly lower speed and large size. In traffic control, signal priority can be given to vehicles classified as bus or an emergency vehicle. The distribution of vehicle types also provides valuable data analysis input to the prediction of highways capacity, assessment of the effectiveness of traffic legislation, automatic toll collection, weight enforcement strategies and environmental impact studies.

As with vehicle detection, a number of technologies were developed for classification. Vision-based, inductive loop, microwave, piezo-electric and acoustic-based classification technologies are the common ones in use nowadays [section 2.1]. Vision-based classification can achieve a correct classification rate higher than 90% [5.1]. The major limitation of vision-based classification is that the system's performance is greatly affected by the environmental and lighting conditions. Classification stations with highly calibrated inductive loops are also in use [5.7] [5.8]. However, the infrastructure and maintenance costs of such a vehicle classification station are high. This makes deployment of such a system economical only at particular sites of interest, such as a toll plaza. On the other hand, vehicle classification by wireless sensor networks provides a much more flexible deployment configuration, making the system portable and, once again, scalable for large scale deployment.

In this chapter, the current classification technologies are first reviewed in section 5.1. The characteristics of magnetic vehicle signatures are studied in section 5.2. The data processing and classification schemes for a platform with limited computation resources are discussed in section 5.3. And the experimental results and analysis are presented in section 5.4.

5.1 Review of Classification Technologies

Vehicle classification refers to the process and methodology to classify a vehicle signature in a specific format into a pre-defined vehicle class (e.g. passenger vehicle or truck). Using the inductive loop system as an example, the inductive signatures and magnetic lengths of a training dataset with known vehicle classes are provided to the classifier for training and calibration. The trained classifier is then used to give the vehicle class estimates of new incoming data samples.

A classification system is evaluated by providing a testing dataset with known vehicle classes to the trained classifier and see how accurate it can classify in each of the classes.

101

The statistical performance is affected by the size of data pool, the distribution of vehicle classes among these dataset, and the classification scheme used. Therefore, correct classification rate weighted by its class sizes is more a representative performance index for comparing different classification systems.

e.g.
$$performance\ index = \frac{1}{n} \sum_{i=1}^{n} \frac{\#\ of\ correctly\ classified\ in\ class\ i}{\#\ of\ class\ i\ samples} \qquad \text{(Eq. 5.1.1)}$$

Various classification schemes are tailored to the needs of different transportation applications. The most well-known is the 13-class Federal Highway Administration (FHWA) scheme shown in Table 5.1.1. It reflects the main concern of most transportation applications: commercial vehicles. The number of axles is an important distinguishing factor in the scheme. Most other schemes are derived by aggregating the FHWA scheme into one with fewer classes and less emphasis on trucks and other commercial vehicles.

Class code	Description
1	Motorcycles
2	Passenger cars
3	Other 4-tire, 2-axle, single-unit vehicles
4	Buses
5	6-tire, 2-axle, single-unit trucks
6	3-axle, single-unit trucks
7	4 or more axle, single-unit trucks
8	4 or less axle, single-trailer trucks
9	5-axle, single-trailer trucks
10	6 or more axle, single-trailer trucks
11	5 or less axle, multi-trailer trucks
12	6-axle, multi-trailer trucks
13	7 or more axle, multi-trailer trucks

Table 5.1.1 13-Class FHWA Classification Scheme

Vision-based, inductive loop, microwave, piezo-electric, and acoustic-based are some of the existing classification technologies in use [section 2.1]. The two most commonly used classification systems, vision-based and inductive loop signature, are discussed in the following sections.

5.1.1 Vehicle Classification by Vision-Based System

Despite the large amount of research on vehicle detection and tracking by vision, there has been relatively little work done in the field of vehicle classification. Given the wide variety of shapes and sizes of vehicles within a single category alone, it is difficult to categorize vehicles using simple parameters. Moreover, the presence of occlusions, shadows, camera noise, changes in lighting and weather conditions are still challenging problems that need to be solved.

Reference	Features	Methodology	Classes	Sample Sizes	Classification Rate
[5.1] Wei, et al.	Ground segmentation, background subtraction, quantization	Artificial neural network	Heavy vehicle, Small car, Motorcycle (3)	Training: 227 Testing: 200	Heavy vehicle (98.5%) Small car (96.92%) Motorcycle (91.94%)
[5.2] S. Gupte, et al.	Vehicle dimensions	Parameters matching	Passenger & Non-passenger vehicles (2)	Training: 100 Testing: 20 min of freeway traffic	90% were tracked, 70% of the tracked vehicles are correctly classified
[5.3] D. Koller, et al.	Parameterized 3D models	Decision functions	Sedan & Hatchback (2)	Training: unspecified Testing: 3	All the 3 presented cases are

					correct
[5.5] G. D. Sullivan, et al.	Wire-frame 3D models	Hypothesis generation, tracking & verification	Passenger & Van (2)	Training: unspecified Testing: 54	96% correct
[5.6] Yuan, et al.	Perspective projections, dimension profiles	k-Nearest Neighbor, heuristics	Passenger, Van, Pickup, Bus, Truck (5)	Training: 50 Testing: 280	91% correct

Table 5.1.1.1 Summary of pervious researches on vision-based vehicle classification

In a simple single camera system, a vehicle may be categorized according to its length and height according to its two dimensional image. In [5.2], such a system is described. The vehicle images are first separated from the background by a segmentation process. In order to tackle the occlusion problem that more than one vehicle are moving on the screen at the same time, individual vehicle tracking is done based on the image sequences. The vehicle parameters are recovered from the 2-D projections of the tracked vehicle. Based on the extracted parameters (mainly length and height), vehicles are being classified. The experiments conducted in [5.2] with such a system shows a successful classification rate of 70% for classifying the vehicles into two classes (passenger and non-passenger).

An object classification approach that uses parameterized three-dimensional models is described in [5.3]. A generic sedan model, represented by a 3D polyhedral model described by 12 length parameters, is used to cover the different shapes of road vehicles. Moreover, the University of Reading has done extensive work in three-dimensional tracking and classification using model matching methods [5.4]. Three-dimensional wire-frame models of various types of vehicles (e.g., sedans, hatchbacks, wagons, etc.) were developed.

Projections of these models were compared to features extracted from the image. Real-time

classification performance is achieved by a simplified version of the model-based tracking

approach that uses orthographic approximations [5.5], a classification rate of 96% is

achieved for classifying 54 vehicles into two classes (passenger and van)[3]. Besides using

the dimensional parameters, other methodologies like k-Nearest Neighbor [5.6] and

artificial neural network [5.1] are also being used. Both experiments in [5.6] and [5.1] show

a classification rate of over 90%.

5.1.2 Vehicle Classification by Inductive Loop Signature

Conventionally, inductive loop detector cards operate in a detection mode, which only

provide the up and down times of the vehicle detection event. However, new detector cards

are being manufactured with the capability to output the raw inductive loop measurements,

known as the vehicle inductive signature. This signature is the result of the net decrease in

the inductance when a vehicle's conducting part passes over the magnetic field generated

by the inductive loop.

Reference	Features	Methodology	Classes	Sample Sizes	Classification Rate
[5.7] Sun, et al.	Lengths, magnitude and moments of inductive signatures	Decision tree	Table 5.1.2.3 (7)	Training: 150 Testing: 150	Overall rate is 90%
		SOFM	Table 5.1.2.4	Training: 26 Testing: 137	80%

[3] Training sample size is unspecified in the reference

105

			(4)		
[5.8] Pursula, et al.	Inductive signatures	SOFM, learning vector, quantization	Table 5.1.2.2 (7)	Training: 663 Testing: 737	96.2%

Table 5.1.2.1 Summary of research on inductive loop-based vehicle classification

Class code	Description
1	Car or van
2	Truck
3	Bus
4	Truck with semi-trailer
5	Truck with trailer
6	Car with trailer
7	Car with mobile home

Table 5.1.2.2 Seven-class scheme used by Pursula and Pikkarainen [5.8]

Vehicle classification was done based on this inductive signature together with other parameters obtained from the inductive loop detector. Pursula and Pikkarainen [5.8] used the seven class scheme in Table 5.1.2.2 for classification using double inductive loop signatures. Inductive loops of the size 3 x 2 meters were used in an array configuration for better accuracy. Self-Organizing Feature Map (SOFM) [5.9] was used with the inductive signature as inputs. A SOFM is an Artificial Neural Network (ANN) that forms clusters of neurons which reflect similarities in the input vector. A correct classification rate of 80% was achieved with the training set. In addition to the SOFM, length windows and Learning Vector Quantization (LVQ) [5.10] were also applied. The resultant correct classification rate was 96.2% out of 737 vehicles.

106

Class code	Description
1	Car, minivan, sports, station, wagon
2	SUV, pickup truck
3	Van, full-size pickup truck
4	Limo
5	2-axle truck
6	Vehicle with trailer, bus
7	Truck with more than 2 axles

Table 5.1.2.3 Seven-class scheme used by Sun [5.7]

Class Code	Description
1	passenger car, minivan, sports car, station wagon
2	SUV, full-size truck, pickup
3	van, limousine
4	cargo truck, vehicle with trailers, bus, and truck with >2 axles

Table 5.1.2.4 Four-class scheme used by Sun [5.7]

In [5.7], Sun had researched the use of existing infrastructure of loop detectors for vehicle classification with two distinct methods. The seven-class scheme shown in Table 5.1.2.4 was used for the first method because it targets at vehicle classes that are not differentiable with current techniques based on axle counting. Its first method uses a heuristic discriminant algorithm for classification and multi-objective optimization for training the heuristic algorithm. Feature vectors obtained by processing inductive signatures are used as inputs into the classification algorithm. Three different heuristic algorithms were developed and yielded an overall classification rate of 90%. Its second method uses Self-Organizing Feature Maps (SOFM) [5.9] with the inductive signature as input. An overall classification

107

rate of 80% was achieved out of 137 samples with the four-class scheme shown in Table 5.1.2.4.

5.2 Magnetic Signature Analysis

As discussed in section 4.4.5, the magnetic signature generated by the magnetic sensor HMC1051Z [3.14] provides much more detail than an inductive loop signature. Such a difference in details level is clearly shown by comparing Fig. 4.2.3.1 and Fig. 4.4.5.1. In this section, a number of characteristics of the magnetic signature that are related to vehicle classification are studied.

5.2.1 Directional Characteristics

The magnetic sensor measures the Earth's magnetic field, which has different inclination angles in different geographical locations. Since the Earth's magnetic field is uniform over the surface in the scale of kilometer, vehicle signatures of the same vehicle measured at adjacent locations in the same traveling direction are expected to be the same. A simple experiment was conducted to confirm this characteristic in Richmond Field Station (RFS) at 08-25-2004 with a Toyota SR5. Two sensor nodes separated by 6 ft were placed in the middle of a lane. The magnetic signatures of the test vehicle from this two sensor nodes show a correlation coefficient of 0.99. This implies that they are virtually the same. A plot of the Z-axis measurements is shown in Fig. 5.2.1.1 below.

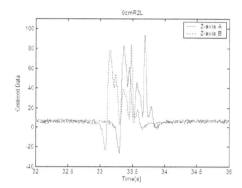

Fig. 5.2.1.1 Z-axis measurements of two nodes with a vehicle traveling in the same direction

However, the Earth's magnetic field line is not perpendicular to the ground surface except at the poles. Therefore, if the same vehicle is being measured when it is traveling in different directions, its magnetic signatures will be slightly different. In principle, with a known magnetic field inclination angle, the magnetic signatures can be normalized with respect to the direction of travel. However, the calculation required is too computationally expensive which is against the implementation objective of this research project. Fortunately, the Earth's magnetic field line angle is ~65 degree from the ground in California [5.11]. This makes the directional difference in signatures not too significant when the difference in traveling direction is small.

Another experiment was conducted to demonstrate this characteristic. Seven sensor nodes were placed side by side along a line perpendicular to the direction of travel. An array of sensor nodes is used to minimize the lateral offset effect in this experiment. Two test vehicles {Toyota Corolla 1989, Ford Taurus 1996} were driven over the sensor nodes in

three different directions {straight, ~20 degree towards right, ~20 degree towards left}.

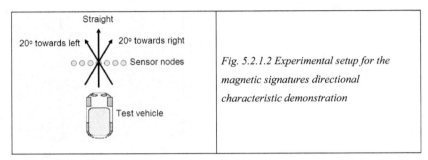

Fig. 5.2.1.2 Experimental setup for the magnetic signatures directional characteristic demonstration

$$Max\ Correlation\ coeff(X * Y * Z) = \underset{sensor\ index\ i,j}{argmax} \{corr(X_i, X_j) \cdot corr(Y_i, Y_j) \cdot corr(Z_i, Z_j)\}$$

(Eq. 5.2.1.1)

The maximum products of the 3-axis correlation coefficients (Eq.5.2.1.1) among the seven

sensor nodes in different runs are shown in Table 5.2.1.1 below. Excepting one case, the

maximum product of correlation coefficients of the same test vehicle with ~20 degree of

directional difference is still higher than 0.9. This makes the signatures acceptable for

classification in California, as long as the difference of their traveling direction is less than

20 degree.

	Run A	Run B	Max Correlation coeff. (X*Y*Z)
Toyota Corolla (1989)	Straight01	Straight02	0.946837292
		Straight03	0.994854323
		Heading Right 01	0.99555328
		Heading Right 02	0.983367722
		Heading Right 03	0.947829884

		Heading Left 01	0.96826485
		Heading Left 02	0.845703555
		Heading Left 03	0.970848939
Ford Taurus (1996)	Straight01	Straight02	0.981689296
		Heading Right 01	0.979513794
		Heading Right 02	0.990692302
		Heading Right 03	0.945619636
		Heading Left 01	0.919551962
		Heading Left 02	0.963439395
		Heading Left 03	0.972325325
	Corolla	Taurus	
Corolla vs Taurus	Straight01	Straight01	0.639722913
		Heading Right 01	0.67434743
		Heading Left 01	0.654170352

Table 5.2.1.1 Maximum of the product of the 3-axis correlation coefficients among the seven sensor nodes in different runs in a directional characteristic experiment

5.2.2 Lateral Offset Characteristics

As one goes from the front to the rear of a vehicle, the measured magnetic field varies significantly, depending on the lateral offset of the measurement. To explore the magnitude of this variation, a set of magnetic measurements was taken for a Ford Taurus 1996 with dimension 16 x 5 ft and 8" clearance. Three-second averaged measurements were taken at the grid positions shown in Fig. 5.2.2.1 below. The corresponding results are shown in Fig. 5.2.2.2.

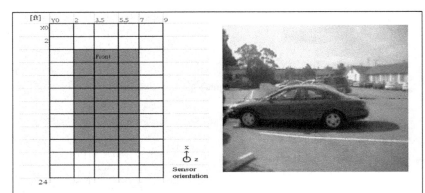

Fig. 5.2.2.1 Experimental layout (left) for the two dimensional magnetic measurement of a Ford Taurus 1996 (right)

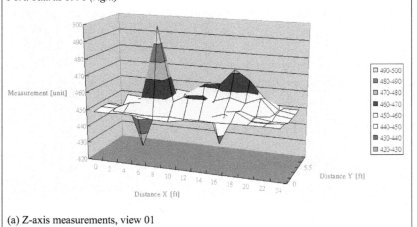

(a) Z-axis measurements, view 01

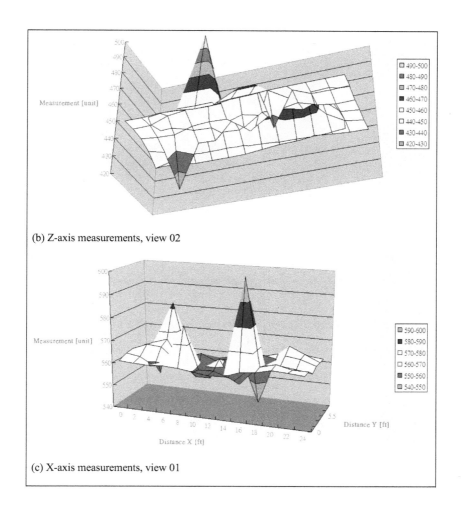

(b) Z-axis measurements, view 02

(c) X-axis measurements, view 01

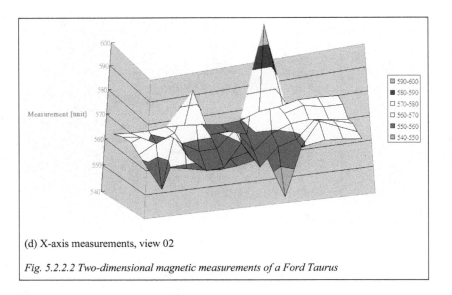

(d) X-axis measurements, view 02

Fig. 5.2.2.2 Two-dimensional magnetic measurements of a Ford Taurus

Let x denote longitudinal and y denote lateral position, with y=0 corresponding to the centerline. From the plots above one may draw two conclusions. First, the magnetic field shows a rough lateral symmetry: the field at (x,y) is similar to the field at (x,-y). The bilateral symmetry is a consequence of the symmetry of the vehicle's mass distribution which is necessary for mechanical balance.

More importantly, however, the magnetic field as a function of x for different lateral offsets y can be completely different. An experiment was designed to demonstrate this characteristic with the setup shown in Fig. 5.2.2.3. Two sensor nodes were placed with a fixed offset distance from the middle of lane in opposite direction. The magnetic signature of a test vehicle {Toyota SR5} was recorded while it was traveling in both directions {right-to-left, left-to-right}.

114

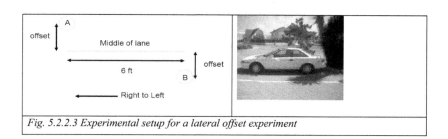

Fig. 5.2.2.3 Experimental setup for a lateral offset experiment

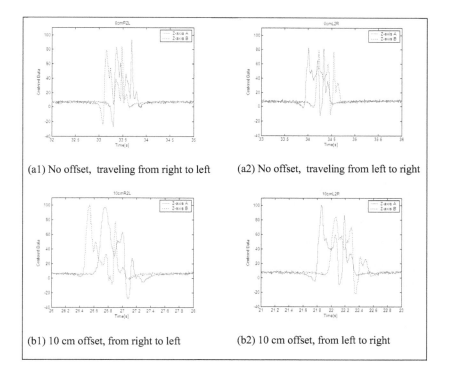

(a1) No offset, traveling from right to left   (a2) No offset, traveling from left to right

(b1) 10 cm offset, from right to left   (b2) 10 cm offset, from left to right

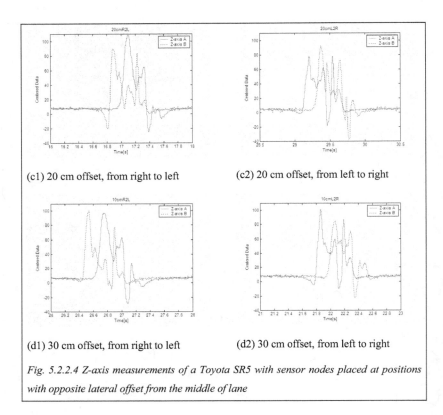

(c1) 20 cm offset, from right to left   (c2) 20 cm offset, from left to right

(d1) 30 cm offset, from right to left   (d2) 30 cm offset, from left to right

Fig. 5.2.2.4 Z-axis measurements of a Toyota SR5 with sensor nodes placed at positions with opposite lateral offset from the middle of lane

The corresponding Z-axis measurements are shown in Fig. 5.2.2.4. It is evident that there is

a large difference between the signatures form nodes A and B even at the small offset of 10

cm. Such a characteristic poses a challenging problem for vehicle classification because it

is difficult to keep a vehicle traveling with the same lateral offset at different sites. A

simple solution would be to use an array of sensor nodes and pick the data pair with the highest correlation coefficient as the sample for classification. More details of this approach are discussed in section 5.3.

5.2.3 Magnetic Signature Examples

In this section, examples of vehicle magnetic signature from different classes of vehicle are presented. For better readability, the smoothed signal as described in section 4.2 was plotted. The Y-axis measurement is omitted as it is usually corrupted by the magnetic signals from vehicles traveling in adjacent lanes. These data were collected at an intersection on Hearst Ave., Berkeley, CA, with the sensor node placed in the middle of the lane. The signatures are plots of magnetic measurements [unit] against time [s] without any normalization. Plots in the same row represent measurements of different magnetic axes.

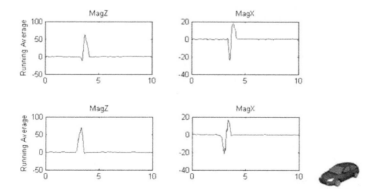

Fig. 5.2.3.1 Magnetic signatures of two Honda passenger vehicles

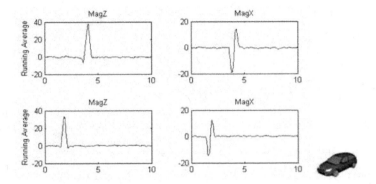

Fig. 5.2.3.2 Magnetic signatures of two Volkswagen passenger vehicles

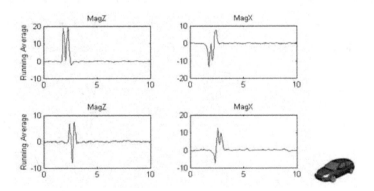

Fig. 5.2.3.3 Magnetic signatures of two Toyota passenger vehicles

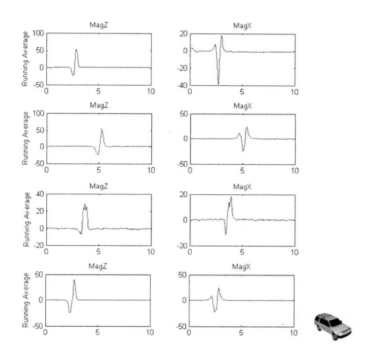

Fig. 5.2.3.4 Magnetic signatures of four SUVs

119

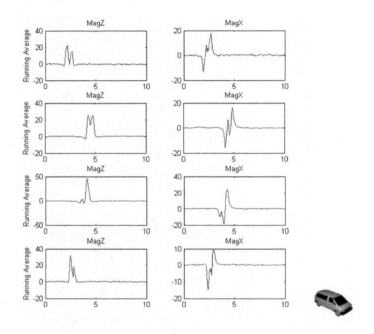

Fig. 5.2.3.5 Magnetic signatures of four Vans

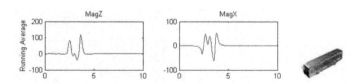

Fig. 5.2.3.6 Magnetic signature of a long bus

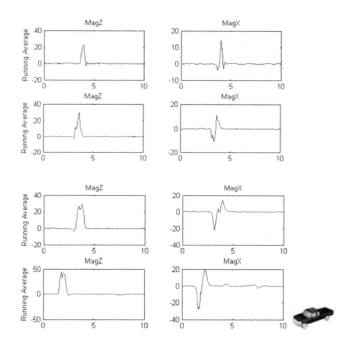

Fig. 5.2.3.7 Magnetic signatures of four pickup trucks

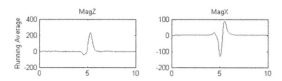

Fig. 5.2.3.8 Magnetic signatures of a two-axle truck

By visual inspection alone one can observe distinguishable magnetic signature patterns among different classes from these examples. The problem is to extract these patterns from the signatures using a classification algorithm that can be executed in real-time and which,

121

at the same time, is simple enough to be implemented within the power constraints of a wireless sensor networks. The proposed data processing and classification schemes are discussed in detail in the following sections.

5.3 Data Processing and Classification Schemes

One can estimate the fraction of traffic due to long vehicles (trucks) and short vehicles (passenger cars) from 30-second average *single* loop measurements of occupancy and counts or volume [5.12][5.13]. The key idea is to this. Recall that

$$Occupancy = flow \times \frac{length}{speed}$$

So, knowing occupancy, flow and speed, one can estimate the average length l. If trucks have a known average length of l_1 and passenger cars have a known average length of l_2, then the fraction p of trucks can be estimated from the relation

$$l = pl_1 + (1-p)l_2.$$

However, this idea will not work if there are three or more vehicle classes or if classification of individual vehicles is needed. In these cases one requires finer measurement. As discussed in section 5.1, conventional classification approaches either involve extensive computation or require a large amount of communication between the sensor nodes and base station, both of which are undesirable for the implementation on a wireless sensor network with its limited computational resources and energy for communication.

With this in mind, new data processing and classification schemas were developed aimed at using the wireless sensor network as a deployment platform. A block diagram of the data processing components for vehicle classification is shown in Fig. 5.3.1 below. The signal "Pre-processing" and "Vehicle Detection" components have already been discussed in Chapter 4.

With the help of the detection flags, the "Vehicle Signature Extraction" component extracts the signature of vehicle i x_i, from a buffer of the raw signal. Because vehicles have unequal lengths and may travel over the magnetic sensor with different speeds, the vehicle signatures will have unequal size (number of samples). The "Transformation" component connects the signature x_i, of size n_i into a data format t_i of fixed vector size N. Next, a principal component analysis is performed to compresses the signature t_i to p_i. Finally, classifiers are applied to generate the classified vehicle class (y_i). Each of these processing steps is discussed in detail in the following sections.

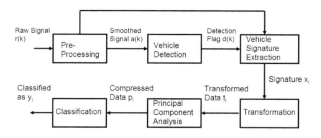

Fig. 5.3.1 Block diagram of the process flow of vehicle classification

5.3.1 Vehicle Signature Extraction

The vehicle signature extraction process is straightforward. The smoothed signal a(k) is first stored in a time-stamped buffer. Once a detection event is committed by the vehicle detection algorithm, the signature of vehicle i x_i, is extracted from the time-stamped buffer with a time window between {up-time – pre-defined extension} and {down-time + pre-defined extension}:

$$x_i = X(i)$$
$$= \{a(k) \, \forall k : (k_{up(i)} - k_{extension}) \leqslant k \leqslant (k_{down(i)} + k_{extension})\}$$

$$\text{(Eq. 5.3.1.1)}$$

The pre-defined extension is a parameter that represents a small time window (e.g. 0.2s) that one would like to include in addition to the detected event time window, in case signals with small amplitude are not caught by the detection events.

5.3.2 Transformation into Average-Bar and Hill-Pattern

The vehicle signatures can have a large and unequal sample vector size n_i. For vehicle length ranging from 5 to 20 meters and speed ranging from 2 to 25 m/s, the detection event duration will range from 0.2 to 10s. With a sampling rate of 128 Hz for each magnetic axis, the resultant sample vector size ranges from 75 to 3840 samples.

Conventionally, the magnitude of the data samples is normalized to 0 mean, magnitude between {-1, 1} [5.28], and re-sampled to a fixed size M (e.g. M = (75+3840)/2 = 1958). However, this step is computationally expensive on the order of the vector size (n_i) since

124

interpolation processes have to be done between all sample points. The resulting vector of size M is still so large that its transmission would consume too much power. According to the analysis in section 5.2, instead of a very detailed signature with all tiny tips, a smooth pattern of the "peak" and "valley" is already good enough for a reasonable classification. Therefore, two transformations, called "Average-Bar" and "Hill-Pattern," are designed to convert the vehicle signatures of variable vector size (n_i) into one with fixed vector size (N):

$$x_i \in R^{n_i} \xrightarrow{\quad AverageBar, HillPattern \quad} t_i \in R^N, N << n_i, \forall i$$

(Eq. 5.3.2.1)

These two transformations are computationally simple enough to be implemented in a sensor node. The resulting small vector size (N) (e.g. 20) makes radio transmission of the transformed data (t_i) feasible from the viewpoint of power consumption. This permits the base station to collect and centralize all the transformed signatures for further classification processing. These two transformation processes are discussed in this section. Their classification performance and experimental results are presented in section 5.4.

5.3.2.1 Average-Bar (AB)

Average-Bar (AB) is a transformation process designed to trim down the vector size (n_i) of vehicle signature (x_i) into one with small and fixed vector size (N), while retaining enough pattern information in the transformed data (t_i) for further classification. The idea is simple. We take a signature vector of size n_i, group the samples into N sub-vectors and replace each of them by its average value. More precisely, the Average-Bar transformation process is:

i, Consider the signature (x_i) with time window K(i)

$$K(i) = \left[k_{start(i)} , k_{end(i)} \right]$$
$$k_{start(i)} = k_{up(i)} - k_{extension} \qquad k_{end(i)} = k_{down(i)} + k_{extension}$$

(Eq. 5.3.2.1.1)

ii, Time window K(i) is divided into N sections

$$\Delta k_{N(i)} = \frac{k_{end(i)} - k_{start(i)}}{N}$$
$$Q(i) = \{ q_{0(i)} \; q_{1(i)} \ldots q_{N(i)} \}$$
$$where \; q_{0(i)} = k_{start(i)}$$
$$q_{j(i)} = q_{(j-1)(i)} + \Delta k_{N(i)} \qquad \forall j: 1 \leqslant j \leqslant N$$

(Eq. 5.3.2.1.2)

iii, The transformed data ($t_{AverageBar(i)}$) with vector size N is given by

$$t_{AverageBar(i)} = B(i) = \{ b_{1(i)} \; b_{2(i)} \ldots b_{N(i)} \}$$
$$b_{j(i)} = avg \left(\{ a(q) \, \forall q : q_{(j-1)(i)} \leqslant q \leqslant q_{j(i)} \} \right) \qquad \forall j: 1 \leqslant j \leqslant N$$

(Eq. 5.3.2.1.3)

Whereas a(q) is a smoothed signal of the signature. Fig. 5.3.2.1.1 below shows three examples of the Average-Bar transformation from the vehicle signature of three 2-axle trucks (FWHA class 5). Signatures of 5 to 8 s and sample vector size 640 to 1024 for each magnetic axis were transformed to a fixed vector size of 20.

126

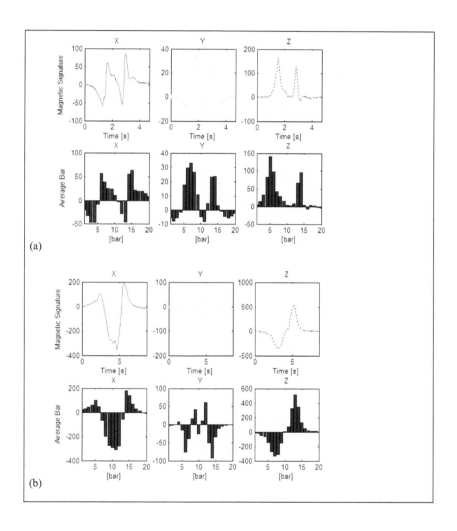

(a)

(b)

127

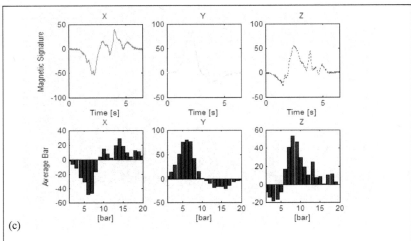

(c)

Fig. 5.3.2.1.1 Three examples of the Average-Bar transformation from vehicle signatures of three 2-axle trucks (FWHA class 5).

iv, Depending on the classification scheme, it can be further normalized to one with magnitude between [-1, 1]

$$\overline{t_{AverageBar(i)}} = \left. B(i) \middle/ \max\left(\left| B(i) \right|\right) \right.$$

(Eq. 5.3.2.1.4)

Fig. 5.3.2.1.2 shows an example of the Average-Bar transformation with magnitude normalization. Observe that the Average-Bar pattern of the same vehicle traveling at different (constant) speeds is the same. Thus speed information of the vehicle is not necessary for Average-Bar processing. This is a significant advantage over other re-sampling techniques that require speed normalization, in which case two sensors are needed to obtain speed and so the re-sampling cannot be done locally within a sensor node.

Secondly, as will be seen in section 5.4, we find that N = 10 or N=20 is sufficient for purposes of classification. This is a significant reduction from the original signatures which can have average size of 1,000 samples.

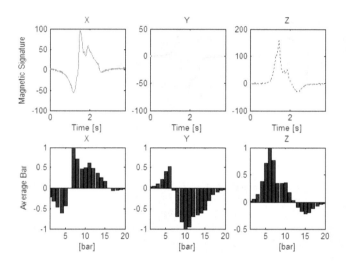

Fig. 5.3.2.1.2 Example of the Average-Bar transformation with normalized magnitude

5.3.2.2 Hill-Pattern

Hill-Pattern (HP) is a transformation process designed to dramatically compress the vehicle signature into a three-valued {+1, 0, -1} signal, while keeping the minimum essential amount of pattern information needed for classification. This numerically simple transformation packs the signal into a highly compressed one, and makes the classification computationally cheap, power efficient and executable in real time. It is very surprising that although almost all of the information in the original signature is dropped by the Hill-

129

Pattern procedure, the classification results remain at a reasonable level with such a simple algorithm. The construction of the Hill-Pattern process is described next while its corresponding classification performance is discussed in section 5.4.

Intuitively, the Hill-Pattern process extracts the pattern of "peaks" and "valleys" exhibited by the smoothed vehicle signature (x_i). A peak is a local maximum and a valley is a local minimum of the signal x_i. Thresholds have to be employed to avoid tiny fluctuations in the signal from contributing to a peak or valley. The precise steps in the Hill-Pattern process are as follows.

i, The rate of change of the signal is first transformed into an intermediate ternary or {+1, 0, -1}-valued signal according to some pre-defined threshold levels

$$c(k) = \frac{\Delta a(k)}{\Delta k}$$

$$\overline{c(k)} = \begin{cases} +1 & c(k) \geqslant c_{+threshold} \\ -1 & c(k) \leqslant c_{-threshold} \\ 0 & Otherwise \end{cases}$$

(Eq. 5.3.2.2.1)

ii, An ad-hoc state machine is used to filter out peaks and valleys (local maxima and minima) with small amplitude or short duration. The state diagram is shown in Fig. 5.3.2.2.1. Unless the amplitude and duration of a hill pattern is large enough, the last hill pattern value will be adopted. Therefore, all the small and short hill pattern can be filtered out.

iii, In the example shown in Fig. 5.3.2.2.2, the resultant hill-pattern 2-bit array is [+1 -1 +1 -1].

130

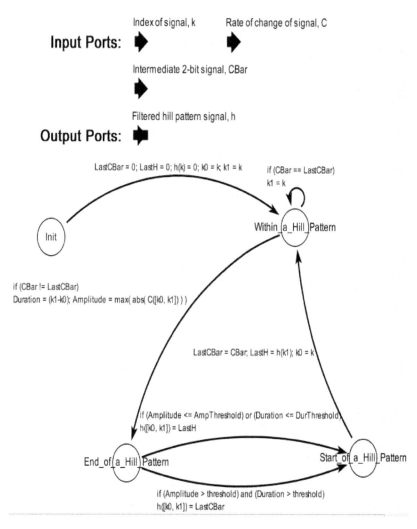

Fig. 5.3.2.2.1 State diagram of the ad-hoc state machine designed to filter out hill patterns with small amplitude or short duration

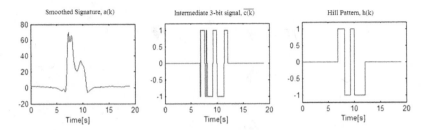

Fig. 5.3.2.2.2 Example of the hill pattern transformation process, [+1 -1 +1 -1]

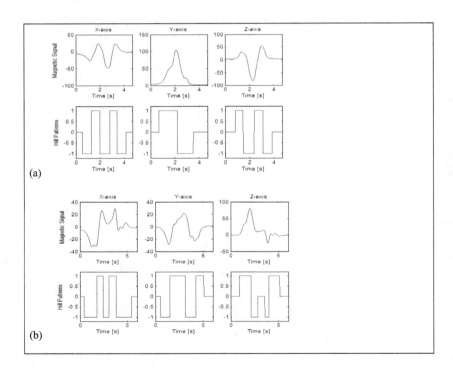

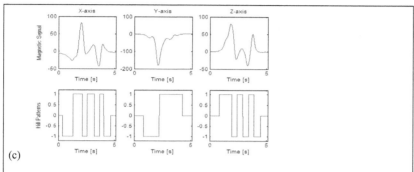

(c)

Fig. 5.3.2.2.3 Three examples of the Average-Bar transformation from vehicle signatures of three 2-axle trucks (FWHA class 5).

5.3.3 Principal Component Analysis

With the Average-Bar and Hill-Pattern transformation processes discussed above, the vehicle signature is now a vector of size as low as 20 for each magnetic axis. However, a combined dimension of 60 may still be too large to be presented directly to a classifier with limited computation power. In view of this, the Principal Component Analysis (PCA) is used to further trim down the signature's dimension.

The PCA, also known as Karhunen-Loeve (KL) method [5.14], provides a principled way to reduce the dimension of the data. It is a linear transformation that transforms data vectors to a new coordinate system such that the greatest variance by any projection of the data comes to lie on the first coordinate, the second greatest variance on the second coordinate, and so on. It is used to reduce the dimension in a dataset while retaining those characteristics of the dataset that most contribute to its variance. It is used frequently in speech data compression and image analysis [5.15]. The objective of PCA is to find a set of

133

n mutually orthogonal vectors $\{e_1 \, e_2 \, ... \, e_n\}$, such that e_1 is the direction in the original data with maximum variance; e2 is the one with the second largest, and so on. This orthogonal set $\{e_1 \, e_2 \, ... \, e_n\}$ is called the set of principal components, and it is used as the basis for the coordinate transformation that gives an alternative representation of the data.

Let X be the matrix whose rows are the row vectors $\{x_1', \, x_2', \, ... \, x_m'\}$ of the target data set, with data size (m) and dimensionality (n). The covariance matrix C of the data is

$$mC = X'X \tag{Eq. 5.3.3.1}$$

The first principal component e_1 is the solution of:

$$\max_{e_1} e_1'Ce_1 \quad with \, \|e_1\|_2 = e_1'Ce_1 \tag{Eq. 5.3.3.2}$$

Similarly, e_2 is obtained by searching for the maximum of $e'Ce$ in the subspace orthogonal to e_1, and so on. The resultant principal components $\{e_1 \, e_2 \, ... \, e_n\}$ are also the eigenvectors of the covariance matrix C with the n largest eigenvalues and the variance in these directions are given by the corresponding eigenvalues. One way to find the principal components is through Singular Value Decomposition (SVD) [5.16]. Recall that the SVD of X is

$$X = USV', \tag{Eq. 5.3.3.3}$$

in which U, V are orthogonal matrices and S is diagonal. The columns of U are called left singular vectors (u_j), the rows of V are called right singular vectors (v_j) and the diagonal values of S are called the singular values (s_j). The ordering of the singular vectors is determined by high-to-low sorting of singular values, with the highest singular value in the upper left index of the S matrix. The covariance matrix C can be written as:

$$mC = X'X = VS^2V' \tag{Eq. 5.3.3.4}$$

134

As a result, the eigenvectors (principal components) and eigenvalues (variances in corresponding direction) of C are given by the right singular vectors v_j and s_j^2. Linear transformation on the data can be done with these principal components, giving:

$$\overline{x_j}' = \begin{bmatrix} x_j'e_1 & x_j'e_2 & \dots & x_j'e_n \end{bmatrix} \quad \forall j : 1 \leqslant j \leqslant m \quad \text{(Eq.5.3.3.5)}$$

This new representation is useful because it allow us to reduce the dimensionality according to the magnitude of the corresponding variances. Suppose the dimension is reduced by projecting the data up to the first J principal components only. The fraction of variance that is retained in the J principal components is

$$\sum_{j=1}^{J} \lambda_j \Big/ \sum_{j=1}^{n} \lambda_j \quad \text{(Eq. 5.3.3.6)}$$

Where λ are the eigenvalues of S. The performance of PCA on the vehicle signature dataset is studied in section 5.4.

5.3.4 Classifiers

Following Fig. 5.3.1, the transformation and PCA processes convert the original signature x_i into the signature p_i. Finally, a classifier makes the classification decision based on p_i. Thus a classifier is a function that assigns to p_i a vehicle class y_i. This assignment may be correct or incorrect. The performance of a classifier is summarized by the fraction of vehicle signatures that are correctly assigned.

Typically, a classifier is selected from a parameterized set of functions. The parameter is set to a value at which the function correctly classifies a set of training signatures. (A training set is simply a set of signatures each of which is labeled with its correct class.)

135

Evidently the larger is the parameterized set of potential classifiers, the larger needs to be the training set. This may lead to a classifier that is computationally intensive.

The proposed solution is to have the classifier running on a powerful computational platform, connected to the base station with simplified signatures (Average-Bar or Hill-Pattern with Principal Component Analysis) collected from the sensor nodes. The algorithms of three classifiers are studied in this section. They are: k-Nearest Neighbor (k-NN), Support Vector Machine (SVM) and Direct Hill-Pattern Matching.

5.3.4.1 k-Nearest Neighbor (k-NN)

k-Nearest Neighbor (k-NN) [5.17][5.18] classification algorithm, is one of the simplest, yet very powerful, classification methods. It is based on the assumption that data samples that are close in the instance space belong to the same class. Therefore, an unseen instance should be classified as the majority class of its k nearest neighbors in the training dataset. The training phase of the algorithm consists in storing the feature vectors and class labels of the training samples. In the actual classification phase, the same features as before are computed for the test sample (whose class is not known). By default, Euclidean distance is used as a measurement of distance between the test sample and the training samples. A summary of some common distance calculation methods is shown in Table 5.3.4.1.1 below.

(a) Euclidean distance	$\sqrt{\sum_i \left(x_{Test,i} - x_{Train,i} \right)^2}$
(b) Sum of absolute differences	$\sum_i \left| x_{Test,i} - x_{Train,i} \right|$
(c) Correlation	$1 - corr\left(x_{Test}, x_{Train} \right)$
(d) Hamming (for binary data)	$\dfrac{number\ of\ different\ bits}{total\ number\ of\ bits}$

Table 5.3.4.1.1 Summary of distance calculation methods for k-NN (where i is the index of the vector elements)

Distances from the new test sample to all samples in the training dataset are computed and the k closest samples are selected. By default, the new point is predicted to belong to the majority class within the set. A summary of some common decision making rules is shown in Table 5.3.4.1.2.

(a) Nearest	Majority rule with nearest point tie-break
(b) Random	Majority rule with random point tie-break
(c) Consensus	Assign "Not classified" unless all the neighbors are in the same class

Table 5.3.4.1.2 Summary of the decision making rules for k-NN

Because of the uneven distribution of number of training sample in each of the vehicle class, a modified decision making rule is used:

i, After calculating the distances between a test sample and all the training samples, k training samples with nearest distance (d) and class (c) are chosen into set K

$$K \in \{(d_1, c_1)\ (d_2, c_2)\ \ldots\ (d_k, c_k)\} \qquad \text{(Eq. 5.3.4.1.1)}$$

ii, Distance (d) is corrected by a factor according to its class (c)

$$\bar{d}_i = d_i \times \left(\frac{\#\ of\ training\ samples\ in\ c_i}{total\ \#\ of\ training\ samples} \right) \quad \forall\ i:1 \leqslant i \leqslant k \qquad \text{(Eq. 5.3.4.1.2)}$$

137

The correction factor increases the distance to classes with a higher proportion in the training samples.

iii, If minimum corrected distance rule is used, the class of the sample with the minimum corrected distance is assigned to the test sample, with majority tie-break

$$j = \underset{i}{\mathrm{minarg}}\, \overline{d}_i \quad \forall i : 1 \leqslant i \leqslant k$$

$$\widehat{c_{Test}} = c_j \qquad\qquad\qquad\qquad\qquad\qquad \text{(Eq. 5.3.4.1.3)}$$

iv, If average corrected distance rule is used, the class with the minimum average corrected distance is assigned to the test sample, with majority tie-break

$$\overline{d}_{c_l} = avg\left(\overline{d}_i \in i : (c_j \equiv c_l)\right) \quad \forall l \in (class\ indexs)$$
$$j = \underset{l}{\mathrm{minarg}}\, \overline{d}_{c_l} \quad \forall l \in (class\ indexs)$$

$$\widehat{c_{Test}} = c_j \qquad\qquad\qquad\qquad\qquad\qquad \text{(Eq. 5.3.4.1.4)}$$

The best choice of k depends upon the data. Generally, a larger value of k reduces the effect of noise on the classification, but makes boundaries between classes less distinct. Theoretically, k can be selected by parameter optimization using cross-validation. The special case that the class is predicted to be the class of the closest training sample (i.e. when k = 1) is called the nearest neighbor algorithm. k = 5 was adopted in the experiments in section 5.4.

Bayesian error rate is the minimum achievable error given the distribution of the data, so it often acts as an evaluation parameter. Suppose, the distribution of x is known, the Bayesian error for classifying a test sample into class C is given by

$$P_{Bayes}(e) = \int \left[1 - P(y = C \mid x)\right] P(x)\, dx$$

The nearest neighbor algorithm has some strong consistency results. As the amount of data approaches infinity, the algorithm is guaranteed to yield an error rate lower bounded by the Bayesian error rate and upper bounded by twice the Bayesian error rate [5.19].

$$P_{bayes}(e) \leqslant P_{k-NN}(e) < 2P_{bayes}(e)$$
(Eq. 5.3.4.1.6)

The k-NN algorithm is easy to implement, but it is computationally intensive, especially when the size of the training dataset grows. Many schemes that seek to reduce the number of distances actually computed have been proposed over the years. Some schemes involve partitioning the feature space, and only computing distances within specific nearby volumes [5.20].

5.3.4.2 Support Vector Machine (SVM)

Maximum-Margin Hyperplane

Support Vector Machine (SVM) was first introduced in 1992 by B.E. Boser [5.21]. It is a set of supervised learning methods used for classification and regression. It became popular because of its success in handwritten digit recognition [5.22]. Its concept is based on a binary linear classification problem. Consider a task of classifying some data ($x \in R^n$) into two classes that can be separated linearly. The problem is to determine whether a hyperplane exists that can separate the data points of that two classes. Fig. 5.3.4.2.1 shows an example of hyperplane in a binary classification problem.

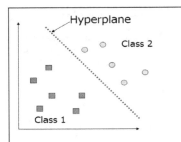 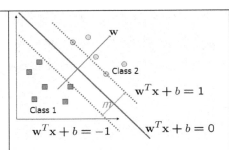

| Fig. 5.3.4.2.1 Example of hyperplane in a binary classification problem | Fig. 5.3.4.2.2 Example demonstrating the maximum-margin hyperplane concept |

The hyperplane should be chosen in such a way that the distance between the closest data points from both classes is maximized. This critical distance is called margin (m), and the corresponding hyperplane is called maximum-margin hyperplane. Fig. 5.3.4.2.2, shows an example demonstrating the maximum-margin hyperplane concept. The maximum-margin hyperplane can be found by the following procedure:

i, Let $\{x_1, \ldots, x_n\}$ be the dataset, and let $y_i \in \{1, -1\}$ be the class label of x_i

ii, The hyperplane should classify all points correctly:

$$y_i\left(w^T x_i + b\right) \geqslant 1 \quad \forall i \qquad \text{(Eq. 5.3.4.2.1)}$$

iii, The margin (m) is given by

$$m = \frac{2}{\|w\|} \qquad \text{(Eq. 5.3.4.2.2)}$$

iv, The maximum-margin hyperplane can be found by solving this constrained optimization problem

$$\min imize \frac{\|w\|^2}{2} \quad subject\ to \quad 1 - y_i\left(w^T x_i + b\right) \leqslant 0 \quad \forall i \qquad \text{(Eq. 5.3.4.2.3)}$$

v, This turns out to be a Quadratic Programming (QP) optimization problem

Lagrangian $\qquad \mathcal{L} = \frac{1}{2}\mathbf{w}^T\mathbf{w} + \sum_{i=1}^{n} \alpha_i \left(1 - y_i(\mathbf{w}^T\mathbf{x}_i + b)\right)$

(Eq. 5.3.4.2.4)

where α is the Lagrange multiplier.

The parameters of the maximum-margin hyperplane are derived by solving this QP optimization problem. There exist several specialized algorithms for quickly solving the QP problem that arises from SVM. The most common method for solving the QP problem is Platt's Sequential Minimal Optimization (SMO) algorithm [5.23].

Soft Margin Hyperplane

A modified maximum margin idea that allows for mislabeled examples is suggested by Corinna Cortes and Vapnik [5.24]. If there exists no hyperplane that can split all the data correctly, the Soft Margin method will choose a hyperplane that splits the data as cleanly as possible, while still maximizing the distance to the nearest cleanly split data. An example demonstrating the soft margin concept is shown in Fig. 5.3.4.2.3, with error measurement ξ_i.

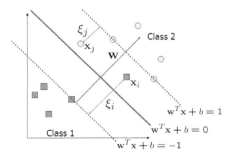

Fig. 5.3.4.2.3 Example demonstrating the soft margin hyperplane concept

The optimization problem becomes

Minimize $\frac{1}{2}||\mathbf{w}||^2 + C \sum_{i=1}^{n} \xi_i$

subject to $y_i(\mathbf{w}^T \mathbf{x}_i + b) \geq 1 - \xi_i, \quad \xi_i \geq 0$

where C is a tradeoff parameter between error and margin

max. $W(\boldsymbol{\alpha}) = \sum_{i=1}^{n} \alpha_i - \frac{1}{2} \sum_{i,j=1}^{n} \alpha_i \alpha_j y_i y_j \mathbf{x}_i^T \mathbf{x}_j$

subject to $C \geq \alpha_i \geq 0, \sum_{i=1}^{n} \alpha_i y_i = 0$

$\mathbf{w} = \sum_{j=1}^{s} \alpha_{t_j} y_{t_j} \mathbf{x}_{t_j}$

(Eq. 5.3.4.2.5)

It is very similar to the hard margin case, except that there is now an upper bound C on α_i.

Non-linear SVM Classifier

An extension of SVM to non-linear classification was suggested by Bernhard Boser, Isabelle Guyon and Vapnik [5.25]. The key idea is to transform the input space ($x \in R^n$) into a feature space ($x \in R^f$) with higher dimension (f > n) by a transformation function $\phi(.)$, so that linear operations can be applied.

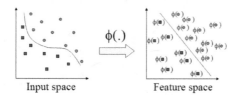

Input space   Feature space

Fig. 5.3.4.2.4 Transformation of input space into the feature space

142

However, the computation in the feature space can be costly because of its high dimension. The Kernel trick is applied to avoid the explicit mapping, by replacing the inner product $x_i^T x_j$ with the kernel function [5.26]

$$K(\mathbf{x}_i, \mathbf{x}_j) = \phi(\mathbf{x}_i)^T \phi(\mathbf{x}_j)$$

(Eq. 5.3.4.2.6)

There are several choices of the kernel function that define different notions of similarity. The one used in this research project is the Radial Basis Function (RBF) kernel (Eq. 5.3.4.2.7). It is popular because it is numerically stable (maps all values to [0 1]) and approximates the linear kernel within certain ranges:

$$K(x_i, x_j) = \exp\left(-\gamma \|x_i - x_j\|^2\right)$$

(Eq. 5.3.4.2.7)

Multi-class SVM Classifier

A direct solution of the multi-class problem using a single SVM formulation is usually avoided because of the complexity. Instead, a combination of several binary SVM classifiers is used to solve a multi-class problem with the one-versus-rest or one-versus-one approach.

Consider a classification problem with (m) classes. The one-versus-rest approach designs a binary classifier for each class, assigning the target class with class index +1 and all the other classes with -1. A new sample is tested against all the m classifiers. If more than one of them gives classification results of +1, it will be assigned to the one with maximum distance from its hyperplane. The one-versus-one approach applies a binary classifier to each pair of the distance classes. A new sample is tested against all m(m-1)/2 classifiers. Each of the classification result counts as one vote for the corresponding class. In the end,

143

the sample is assigned to the class with maximum number of votes. According to [5.27], the performance of these two approaches is competitive.

5.3.4.3 Direct Hill-Pattern Matching

This computationally extremely simple classifier is designed in such a way as to implement the whole classification process within a sensor node. Again, there is a tradeoff between classification performance and simplicity of the algorithm. The main data input is the 3-bit hill-pattern array discussed in section 5.3.2.2. It is virtually a table-looks-up classifier once the training phase is done. The algorithm is as follows:

i, With a training dataset of N classes comprising m samples $\{(h_{x1}, h_{y1}, h_{z1}, c_1), \ldots (h_{xm}, h_{ym}, h_{zm}, c_m)\}$, the counts of a specific hill-pattern array belonging to each class are recorded: h_{xi}, h_{yi}, h_{zi} are the hill-pattern arrays of X, Y and Z-axis respectively, and c_i is the corresponding correct class.

$$Count(h_{ai,}c_i) = Count(h_{ai,}c_i) + 1 \quad \forall a \in \{x, y, z\} \quad \forall i : 1 \leqslant i \leqslant m$$

(Eq. 5.3.4.3.1)

ii, A probability distribution of different classes of each hill-pattern array is calculated according to the counts in step (i).

$$Prob(h_{aj,}c_k) = \frac{Count(h_{aj,}c_k)}{\sum_l Count(h_{aj,}c_l)} \quad \forall a \in \{x, y, z\} \quad \forall j, k$$

(Eq. 5.3.4.3.2)

iii, When a set of testing hill-pattern arrays $\{h_{xi}, h_{yi}, h_{zi}\}$ is given, a combined probability distribution is calculated according to the product of its probability distributions in different axes. And the class with the maximum combined probability will be assigned to the test sample.

$$Prob(h_{xi}, h_{yi}, h_{zi}, c_k) = Prob(h_{xi}, c_k) \times Prob(h_{yi}, c_k) \times Prob(h_{zi}, c_k) \quad \forall k$$

$$c_i = \underset{ck}{\text{maxarg}} \{ Prob(h_{xi}, h_{yi}, h_{zi}, c_k) \} \quad \forall k$$

(Eq. 5.3.4.3.3)

iv, In case a new hill-pattern array (h_{new}) is presented, it will be fitted to its closest hill-pattern array according the following algorithm:

$$Length\ factor \quad f_L = |length(h_i) - length(h_{new})|$$

($h_{new,I}$) and ($h_{i,I}$) are constructed by adding elements to the shorter array in a way that the number of matching elements is maximized.

$$Insertion\ factor \quad f_I = length(h_{new,I}) - count\_bit(h_{i,I}\ \&\ h_{new,I})$$

($h_{new,D}$) and ($h_{i,D}$) are constructed by deleting elements from the longer array in a way that the number of matching elements is maximized.

$$Deletion\ factor \quad f_D = length(h_{new,D}) - count\_bit(h_{i,D}\ \&\ h_{new,D})$$

The closest hill-pattern array is identified as the one with a minimum overall factor.

$$Overall\ factor \quad f_{i,overall} = f_{i,L} + f_{i,I}/2 + f_{i,D}/2$$

(Eq. 5.3.4.3.4)

The performance of this simple classifier counts on a rich training dataset to provide a good enough hill-pattern arrays statistics database. It may not be the best in terms of correct classification rate, but it is simple enough to be implemented by the microprocessor in the sensor node in real time. Its experimental results and analysis are presented in the next section.

5.4 Experimental Results and Analysis

A number of experiments focusing on vehicle classification are discussed in this section: 5.4.1 discusses the FHWA class scheme used in the classification experiments. 5.4.2 (Dataset C1) presents the results of a preliminary experiment aimed at testing the feasibility of using Hill-Pattern for classification. 5.4.3 (Dataset C2) presents the classification results of 256 trucks using Hill-Pattern with a direct matching classifier. 5.4.4 (Dataset C3) presents the classification results of 839 vehicles with different combination of transformations and classifiers. 5.4.5 (Dataset C4) presents the classification results of 864 trucks using magnetic lengths for a two-level classification.

5.4.1 FHWA Classification Schemes

Various classification schemes are tailored to the needs of different transportation applications. The most well-known is the 13-class Federal Highway Administration (FHWA) scheme shown in Table 5.4.1.1. It reflects the main concern of most transportation applications: commercial vehicles. The number of axles is an important distinguishing factor in the scheme. Most other schemes are derived by aggregating the FHWA scheme into fewer classes, with less emphasis on trucks and other commercial vehicles.

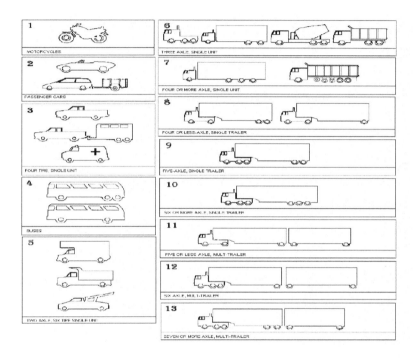

Table 5.4.1.1 13-classes FHWA classification scheme

The classification of trucks is of particular interest in many applications. Trucks and other

oversized vehicles have distinctly different performance characteristics from passenger

vehicles. They travel at a slower speeds, occupy more road space, induce more damage to

the pavements, require longer braking distance and time, and are sometimes lane-restricted.

In terms of traffic flow considerations, an accurate measurement of trucks on the road will

lead to more accurate modeling and simulation of the traffic. In terms of pavement

management, the truck information is useful in the design of new pavement and the life time analysis of the existing pavement. Therefore, the classification experiments discussed below is mainly focused on trucks.

5.4.2 Preliminary Hill-Pattern Classification (Dataset C1)

This section presents the results of a preliminary Hill-Pattern classification experiment. The objective is to study the feasibility of using the simple Hill-Pattern arrays [section 5.3.2.2] in classification. The magnetic signatures of 37 vehicles (Dataset C1) were collected by a single dual-axis (Z and X) sensor node, sampling at 64 Hz for each of the axis. The distribution of vehicle classes in dataset C1 is shown in Table 5.4.2.1. Hill-Patterns of signatures were generated as discussed in section 5.3.2.2. And the direct hill-pattern matching classifier discussed in section 5.3.4.3 is used for classification.

Class	Passenger (P)	SUV	Van	Pickup-Truck	Bus
Observed	15	7	5	7	3

Table 5.4.2.1 Distribution of vehicle classes in dataset C1

Fig. 5.4.2.1 displays the signatures and hill patterns of four passenger vehicles (P). There are six plots per vehicle. The top row shows the magnetic signatures of Z and X-axis (along direction of travel). The second row shows the corresponding hill patterns. All the passenger vehicle cases shown in Fig. 5.4.2.1 give a single "peak" pattern in the Z-axis measurements, resulting in a hill-pattern array of $\{+1, -1\}$. Similarly, a "valley" followed by a "peak" pattern is shown in the X-axis measurements, resulting in a hill-pattern array of $\{-1, +1, -1\}$. These extracted common hill-pattern arrays were used with the direct

148

matching classifier. The third row gives the outcome of the classifier, which classifies the signature into seven classes defined in Table 5.4.2.2. All the cases in Fig. 5.4.2.1 are correctly classified.

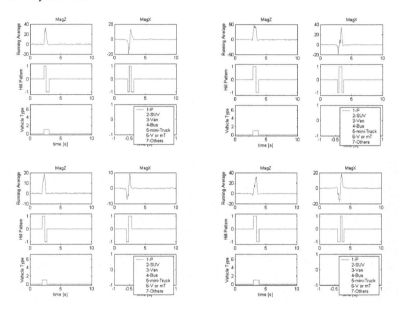

Fig. 5.4.2.1 Vehicle signatures and hill-patterns of four passenger vehicles in dataset C1

Fig. 5.4.2.2 displays the signatures, hill patterns and classifications of four SUVs. Except for the bottom left case, the Z and X-axis hill-pattern arrays of all the other three cases are different from those of passenger vehicles. These hill-pattern arrays are {-1. +1. -1} and {+1, -1, +1, -1} respectively. However, the bottom left case is misclassified as a passenger vehicle, because its initial negative slope of the Z-axis measurements is too small in magnitude to cross the threshold, and an initial positive slope in X-axis measurements is missing. It is possible that this SUV has a different build from others, making its signature

149

similar to that of a passenger vehicle. If this kind of SUV is longer than passenger vehicles

in general, using magnetic length as a classification feature might lead to a correct

classification.

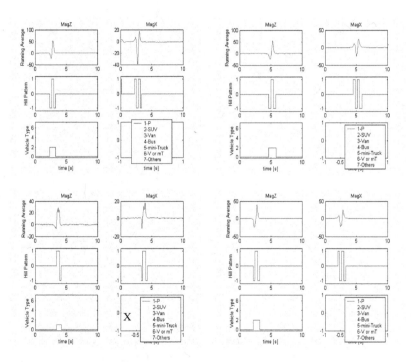

Fig. 5.4.2.2 Vehicle signatures and hill-patterns of four SUVs in dataset C1

Fig. 5.4.2.3 displays the signatures, hill patterns and classifications of four Vans. Three out

of four cases are correctly classified. The common hill-pattern arrays of Vans are {+1, -1,

+1, -1} and {-1, +1, -1} for Z and X-axis respectively. The Z-axis measurements of the misclassified case do have two peaks, but the magnitude of the first one is too small to cross the threshold. Again, incorporation of magnetic length might have helped.

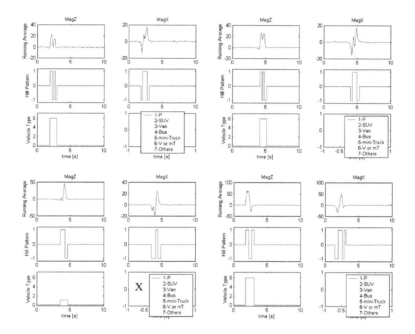

Fig. 5.4.2.3 Vehicle signatures and hill-patterns of four Vans in dataset C1

The confusion matrix of different classification schemes with full dataset training is shown in Table 5.4.2.2. and 5.4.2.3. Using the 7-classes scheme defined in Table 5.4.2.2, 24 out of 37 vehicles (63%) are correctly classified. A FHWA-based scheme is indicated in Table 5.4.2.3, with passenger vehicles and SUVs belong to FHWA class 2, Vans and pickup truck belong to FHWA class 3 and buses belongs to FHWA class 4. 31 out of 37 vehicles (83%) are correctly classified in terms of this FHWA-based scheme.

Observed		Classified as					
		P	SUV	Van	Bus	Pickup	Van or Pickup
P	15	11	4				
SUV	7	3	4				
Van	5	1		1			3
Bus	3				3		
Pickup	7	4	1			2	

Table 5.4.2.2 Confusion matrix of dataset C1 classification with a 7-class scheme

Observed		Classified as		
		FWHA-2	FWHA-3	FWHA-4
FWHA-2 (P+SUV)	22	22		
FWHA-3 (Van + Pickup)	12	6	6	
FWHA-4 (Bus)	3			3

Table 5.4.2.3 Confusion matrix of dataset C1 classification with a FHWA based scheme

Although the sample size of this dataset (C1) is too small to make any firm judgment, promising results were shown by the hill-pattern technique. A reasonable classification rate is achieved by using a simple classifier based on the signatures from a *single* sensor node. Even though the data in this experiment were generated off-line, the real time implementation of such a simple algorithm is quite feasible.

5.4.3 Hill-Pattern Classification of 256 Trucks (Dataset C2)

As discussed in section 5.4.1, the classification of trucks is of special interest in many applications. A four-hour trace of 3-axis magnetic measurements sampling at 128Hz for each axis was recorded at a weigh-in-motion station on I-880 in Fremont, CA on Dec 21st, 2004. The signatures of 256 trucks were collected with speed and magnetic length estimations obtained by a sensor node pair as described in chapter 4. Ground truth of the vehicle classes was extracted manually from the recorded video. A total of five different classes of truck were found in dataset C2. The distribution of vehicle classes and estimated magnetic lengths is shown in Table 5.4.3.1.

FWHA Class	Counts		Estimated Magnetic Length [m]			
			Mean	Min	Max	Std
5	131		11.11	6.29	30.54	3.23
6	14		12.48	8.69	21.94	3.34
8	30		18.48	12.02	25.81	4.53
9	68		22.35	7.01	56.41	6.32
11	22		21.04	5.64	29.48	5.53

Table 5.4.3.1 Distribution of vehicle classes and estimated magnetic lengths of dataset C2

Hill-Patterns of signatures were generated as discussed in section 5.3.2.2. And the Direct Hill-Pattern Matching classifier discussed in section 5.3.4.3 was used for classification. Fig. 5.4.3.1 displays the signatures and hill patterns of two FHWA class-5 trucks, and Fig. 5.4.3.2 displays those of two FHWA class-9 trucks. There are six plots per vehicle. The left column shows the magnetic signals from each of the three axes. The right column shows the corresponding hill-patterns.

153

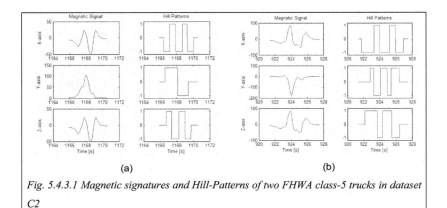

Fig. 5.4.3.1 Magnetic signatures and Hill-Patterns of two FHWA class-5 trucks in dataset C2

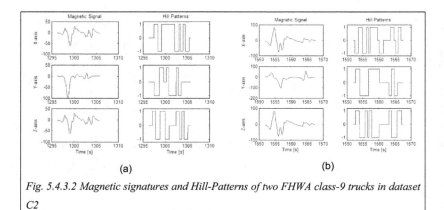

Fig. 5.4.3.2 Magnetic signatures and Hill-Patterns of two FHWA class-9 trucks in dataset C2

Table 5.4.3.2 shows the confusion matrix of hill-pattern classification with full dataset training from all the three-axis data. A correct classification is declared if a test sample is classified as its observed class. Surprisingly, even if the magnetic length is not used, a classification rate of 80% is reached. Table 5.4.3.3 shows the correct classification rates

154

when different features are applied. When the magnetic length is used to separate samples between FHWA classes {5, 6} from {8, 9, 11}, the classification rate is boosted to 82.64%. However, when only 200 samples are used for training and 65 for testing, the classification rate drops to 61.5%. When 165 samples are used for training and 100 for testing, the classification rate is only 54.0%. The degraded performance is expected since the performance of this simple direct matching classifier depends heavily on a rich set of statistics from the training database. The classification rate of Hill-Pattern Classification is expected to reach 80% if a large and evenly distributed training dataset is provided.

[Counts]	Classified as					Sum	Correct %
Observed	5	6	8	9	11		
FWHA Index							
5	95	26	8	2	0	131	72.52
6	2	12	0	0	0	14	85.71
8	2	3	25	0	0	30	83.33
9	1	0	4	60	3	68	88.24
11	0	0	2	0	20	22	90.91
					Overall	265	80.00

[%]	Classified as				
Observed	5	6	8	9	11
FWHA Index					
5	72.52	19.85	6.11	1.53	0.00
6	14.29	85.71	0.00	0.00	0.00
8	6.67	10.00	83.33	0.00	0.00
9	1.47	0.00	5.88	88.24	4.41
11	0.00	0.00	9.09	0.00	90.91

Table 5.4.3.2 Confusion matrix of the Hill-Pattern Classification for dataset C2 without using magnetic length

Features	Overall Correct Classification %
Hill-Pattern X and Magnetic Length	53.21
Hill-Pattern Y and Magnetic Length	62.64
Hill-Pattern Z and Magnetic Length	59.25
Hill-Pattern XYZ	80.00
Hill-Pattern XYZ and Magnetic Length	82.64
Hill-Pattern XYZ with 200 samples for training, 65 for testing	61.50
Hill-Pattern XYZ with 165 samples for training, 100 for testing	54.00

Table 5.4.3.3 Overall correct classification rates of Hill-Pattern Classification with different features applied on dataset C2

5.4.4 k-NN and SVM Classification of 839 Vehicles (Dataset C3)

In this experiment, another set of magnetic measurements sampling at 128Hz was recorded at the same weigh-in-motion station as the one in section 5.4.3. A total of 839 vehicle signatures were extracted from dataset C3. Again, ground truth of the vehicle classes was extracted manually from the recorded video. The distribution of vehicle classes in dataset C3 is shown in Table 5.4.4.1.

FHWA Class	2	5	6	8	9	11		Total
#	85	359	45	83	194	73		839

Table 5.4.4.1 Distribution of vehicle classes in dataset C3

The objective of this experiment is to compare the performance of different classifiers with different transformations on data. Three different binary classification schemes were used in order to make the results clearer and easily comparable. Both the Average-Bar and Hill-Pattern transformations were applied to this dataset with a vector size of 20 for each of the

156

axis. The three-axis samples were stacked into a long vector with size of 60. Principal

Component Analysis (PCA) was applied to the transformed data with three different

variance drops {0, 0.001, 0.01}. The corresponding dimensions of the average-bar and hill-

pattern data were reduced to {60, 58, 40} and {60, 35, 21} respectively. 50% of samples in

each of these classes were randomly picked from this compressed dataset for training. The

reminding 50% of samples were used as test data. This cross-validated dataset was passed

to the k-NN and SVM classifiers as discussed in section 5.3. 12 different sets of

classification result were generated from these dataset and classifier combinations. This

whole cross-validation and classification process was repeated 10 times and the averaged

results are presented in the following section.

5.4.4.1 Classification of FHWA Class-2 against 5, 6, 8, 9, 11

Due to the loss of X and Y-axis magnetic data in some of the passenger vehicle samples,

only Z-axis data were used for classifying FHWA class-2 against 5, 6, 8, 9, 11. The

distribution of training and testing data is shown in Table 5.4.4.1.1. A summary of the

classification results is shown in Table 5.4.4.1.2. A maximum classification rate of 0.97

was achieved by using k-NN with k = 5 on the Average-Bar data without PCA

compression. The corresponding confusion matrix is shown in Table 5.4.4.1.3.

	FWHA 2	FWHA 5 6 8 9 11
Training	43	377
Test	42	377
Total	85	754

Table 5.4.4.1.1 Distribution of data for classifying FHWA class-2 against 5, 6, 8, 9, 11

	Variance drop in PCA		
	0	0.001	0.01
SVM on Average-Bar Data			
Reduced Dimension	20.00	20.00	13.00
Average Correct Rate	0.87	0.87	0.80
Training Correct Rate	1.00	1.00	1.00
k-NN on Average-Bar Data			
Reduced Dimension	20.00	20.00	13.00
Average Correct Rate	0.97	0.97	0.97
Training Correct Rate	0.98	0.98	0.98
SVM on Hill-Pattern Data			
Reduced Dimension	20.00	16.00	11.00
Average Correct Rate	0.88	0.88	0.88
Training Correct Rate	0.92	0.92	0.92
k-NN on Hill-Pattern Data			
Reduced Dimension	20.00	16.00	12.00
Average Correct Rate	0.91	0.91	0.91
Training Correct Rate	0.91	0.91	0.91

Table 5.4.4.1.2 Summary of results for classifying FHWA class-2 against 5, 6, 8, 9, 11 using Z-axis data only

[%]	Classified as		Counts
Observed Class	2	5,6,8,9,11	
2	0.97	0.03	42
5,6,8,9,11	0.03	0.97	377
Avg. Counts	52.3	366.7	

Table 5.4.4.1.3 Confusion matrix of k-NN on Average-Bar data for classifying FHWA class-2 against 5, 6, 8, 9, 11

5.4.4.2 Classification of FHWA Class-5 against 6, 8, 9, 11

The three-axis magnetic data were used for classifying FHWA class-5 against 6, 8, 9, 11. The distribution of training and testing data is shown in Table 5.4.4.2.1. A summary of the classification results is shown in Table 5.4.4.2.2. A maximum classification rate of 0.80 was achieved by using k-NN on the Average-Bar data with the three different PCA variance drops {0, 0.001, 0.01}. The confusion matrix for the 0.01 variance case drop is shown in Table 5.4.4.2.3.

	FWHA 5	FWHA 6 8 9 11
Training	180	198
Test	179	197
Total	359	395

Table 5.4.4.2.1 Distribution of data for classifying FHWA class-5 against 6, 8, 9, 11

	Variance drop in PCA		
	0	0.001	0.01
SVM on Average-Bar Data			
Reduced Dimension	60.00	60.00	40.00
Average Correct Rate	0.75	0.74	0.74
Training Correct Rate	1.00	1.00	1.00
k-NN on Average-Bar Data			
Reduced Dimension	60.00	60.00	40.00
Average Correct Rate	0.80	0.80	0.80
Training Correct Rate	0.86	0.87	0.87
SVM on Hill-Pattern Data			
Reduced Dimension	60.00	35.00	21.00

159

Average Correct Rate	0.69	0.67	0.65
Training Correct Rate	0.95	0.95	0.95
k-NN on Hill-Pattern Data			
Reduced Dimension	60.00	36.00	21.00
Average Correct Rate	0.70	0.70	0.71
Training Correct Rate	0.78	0.79	0.80

Table 5.4.4.2.2 Summary of results for classifying FHWA class-5 against 6, 8, 9, 11

[%]	Classified as		Counts
Observed Class	5	6,8,9,11	
5	0.91	0.09	179
6,8,9,11	0.30	0.70	197
Avg. Counts	222.2	153.8	

Table 5.4.4.2.3 Confusion matrix of k-NN on Average-Bar data with 0.01 variance drop of PCA for classifying FHWA class-5 against 6, 8, 9, 11

5.4.4.3 Classification of FHWA Class-9 against 6, 8, 11

The three-axis magnetic data were used for classifying FHWA class-9 against 6, 8, 11. The distribution of training and testing data is shown in Table 5.4.4.3.1. A summary of the classification results is shown in Table 5.4.4.3.2. A maximum classification rate of 0.71 was achieved by using k-NN on the Average-Bar data with PCA of 0.01 variance drops. The confusion matrix of it is shown in Table 5.4.4.3.3.

	FWHA 9	FWHA 6 8 11
Training	97	101
Test	97	100
Total	194	201

Table 5.4.4.3.1 Distribution of data for classifying FHWA class-9 against 6, 8, 11

	Variance drop in PCA		
	0	0.001	0.01
SVM on Average-Bar Data			
Reduced Dimension	60.00	60.00	40.00
Average Correct Rate	0.66	0.65	0.65
Training Correct Rate	1.00	1.00	1.00
k-NN on Average-Bar Data			
Reduced Dimension	60.00	60.00	43.00
Average Correct Rate	0.69	0.68	0.71
Training Correct Rate	0.79	0.78	0.78
SVM on Hill-Pattern Data			
Reduced Dimension	60.00	37.00	22.00
Average Correct Rate	0.57	0.58	0.58
Training Correct Rate	0.99	0.99	0.99
k-NN on Hill-Pattern Data			
Reduced Dimension	60.00	36.00	22.00
Average Correct Rate	0.57	0.57	0.58
Training Correct Rate	0.70	0.70	0.71

Table 5.4.4.3.2 Summary of results for classifying FHWA class-9 against 6, 8, 11

[%]	Classified as		Counts
Observed Class	9	6,8,11	
9	0.58	0.42	97
6, 8, 11	0.22	0.78	100
Avg. Counts	77.9	119.1	

Table 5.4.4.3.3 Confusion matrix of k-NN on Average-Bar data with 0.01 variance drop of PCA for classifying FHWA class-9 against 6, 8, 11

To summarize, datasets with Average-Bar transformation give roughly 10% better classification rate than that with Hill-Pattern. Even through a large portion of the original information is dropped in this 3-bit Hill-Pattern signal, a reasonable level of classification rate was achieved {0.91, 0.71, 0.58} in different datasets. The results indicate that this simple and efficient transformation algorithm has a high potential in applications which need a highly compressed vehicle magnetic signature.

For datasets with Average-Bar transformation, the k-NN classifier gives a 3-10% better classification rate than that using SVM. The performances of k-NN and SVM are close for datasets with Hill-Pattern transformation. This is due to the large variation in different vehicle signatures, making the nonlinearity of the datasets high and degrading the performance of a SVM classifier.

5.4.5 Classification of 864 Trucks with Magnetic Length (Dataset C4)

In this experiment, another set of magnetic measurements sampling at 128Hz was recorded at the same weigh-in-motion station as in section 5.4.3. A total of 864 truck signatures were extracted from dataset C4 with speed and magnetic length estimates. Ground truth of the

vehicle classes was extracted manually from the recorded video. The distribution of vehicle

classes in dataset C4 is shown in Table 5.4.5.1.

FWHA Class	5	6	8	9	11	Total
#	435	47	106	236	40	864

Table 5.4.5.1 Distribution of vehicle classes in dataset C4

The objective of this experiment is to test the use of magnetic length in classifying trucks.

Two sets of classification processes were applied to dataset C4. The first scheme used the

same techniques as discussed in section 5.4.4 without using magnetic length estimates. The

second scheme used two levels of classification. The first level used magnetic length

estimates to separate FHWA class {5, 6} from {8, 9, 11}. The second level further

classified the separated groups into its own class. The results are presented in the following

section.

5.4.5.1 Single-Level Classification W/O Using Magnetic Length

Similar to section 5.4.4, different transformations and classifiers were used with the dataset.

A variance drop of 0.01 in PCA (dimension reduced from 60 to 40) was used throughout

this experiment. The classification results are summarized in Table 5.4.5.1.1.

	FHWA Classes	
	5 vs 6 vs 8 vs 9 vs 11	5 6 vs 8 9 11
SVM on Average-Bar Data		
Average Correct Rate	0.41	0.68
Training Correct Rate	1.00	1.00
k-NN on Average-Bar Data		

Average Correct Rate	0.64	0.85
Training Correct Rate	0.81	0.89
SVM on Hill-Pattern Data		
Average Correct Rate	0.48	0.72
Training Correct Rate	0.94	0.97
k-NN on Hill-Pattern Data		
Average Correct Rate	0.60	0.77
Training Correct Rate	0.72	0.83

Table 5.4.5.1.1 Summary of classification results of a single-level classification on dataset C4 without using magnetic length.

[%]	Classified as					Counts
Observed Class	5	6	8	9	11	
5	0.84	0.07	0.05	0.03	0.01	217
6	0.60	0.23	0.06	0.11	0.00	23
8	0.42	0.03	0.22	0.29	0.04	53
9	0.18	0.03	0.17	0.55	0.08	118
11	0.07	0.02	0.07	0.36	0.49	20
Avg. Counts	241.6	24.7	44.5	97.6	22.6	

Table 5.4.5.1.2 Confusion matrix of k-NN on Average-Bar data of dataset C4 for classifying FHWA class {5 vs 6 vs 8 vs 9 vs 11} without using magnetic length

For the classification of FHWA classes {5 vs 6 vs 8 vs 9 vs 11}, the highest classification rate is 0.64 using k-NN on the Average-Bar data. The corresponding confusion matrix is shown in Table 5.4.5.1.2. The main confusion is in the classification of class {5 vs 6 vs 8} and {9 vs 11}. 60% and 42% of class 6 and 8 samples respectively are classified as class 5; and 36% of class 11 samples are classified as class 9. This is caused by the small sample sizes of classes 6 and 11 (see Table 5.4.5.1). Even through a correction factor

$(1 - \frac{num\ of\ class\ i\ samples}{total\ num\ of\ samples})$ was introduced to account for the unbalanced dataset, the database is

still under-trained in those classes with a small sample size.

5.4.5.2 Two Levels of Classification with the Use of Magnetic Length

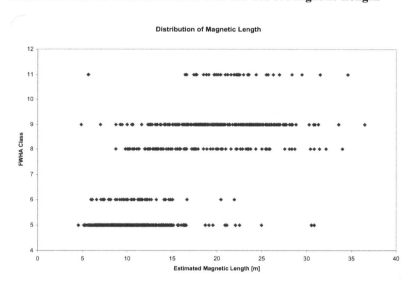

Fig. 5.4.5.2.1 Distribution of estimated magnetic lengths of dataset C4

	FHWA Classes				
Magnetic Length [m]	**5**	**6**	**8**	**9**	**11**
Mean	10.35	11.22	17.89	20.79	21.43
Std	3.21	3.36	6.31	6.11	5.41

Table 5.4.5.2.1 Statistic of estimated magnetic lengths of dataset C4

[%]	Classified as		Counts
Observed Class	**5,6**	**8, 9, 11**	
5,6	0.94	0.06	241
8, 9, 11	0.20	0.80	191
Avg. Counts	264.3	167.7	

Table 5.4.5.2.2 Confusion matrix of using estimated magnetic lengths of dataset C4 for classifying FHWA class {5, 6} vs {8, 9, 11}

The first level of classification used the estimated magnetic lengths to separate FHWA class {5, 6} from {8, 9, 11}. The distribution of estimated magnetic lengths of dataset C4 is shown in Fig. 5.4.5.2.1., and summary statistics are given in Table 5.4.5.2.1. According to the statistical analysis, a cut-off length of $L_{cutoff} = 15.4$[m] was used to separation the two groups.

$$Group_{L1}(i) = \begin{cases} \{5,6\} & if \ Magnetic\_Length(i) \leq L_{cutoff} \\ \{8,9,11\} & Otherwise \end{cases} \quad \forall i$$

(Eq.5.4.5.2.1)

A classification rate of 0.88 was achieved by magnetic length alone for this first level. The confusion matrix is shown in Table 5.4.5.2.2.

The second level further classified the groups from the first level into separated class with Average-Bar transformation and k-NN classifier. The confusion matrix of this second level classification is presented in Table 5.4.5.2.3 below.

[%]	Classified as			Counts
Observed Class	**5**	**6**	**Others**	
5	0.92	0.08	0.00	204.6
6	0.85	0.15	0.00	22.3
Others	0.91	0.09	0.00	37.4

Avg. Counts	242.1	22.2	0	

[%]	Classified as				Counts
Observed Class	8	9	11	Others	
8	0.43	0.50	0.07	0.00	30.3
9	0.24	0.64	0.12	0.00	104.7
11	0.13	0.40	0.47	0.00	18.6
Others	0.42	0.49	0.09	0.00	14.1
Avg. Counts	46.2	96.3	25.2	0	

Table 5.4.5.2.3 Confusion matrix of the second level classification of dataset C4 for classifying FHWA class {5 vs 6} and {8 vs 9 vs 11}

Avg. Classification Rate	FWHA Classes					
	5	6	8	9	11	Overall
Single-Level (5.4.5.1)	0.84	0.23	0.22	0.55	0.49	0.64
Two-Level (5.4.5.2)	0.92	0.15	0.43	0.64	0.47	0.65

Table 5.4.5.2.4 Comparison of the classification rates between single-level and two-level classification in dataset C4

A comparison of the classification rates between the single-level and two-level classification is shown in Table 5.4.5.2.4. The under-trained dataset problem discussed in section 5.4.5.1 also arises in this two-level classification. Because of this defect in the dataset, the use of magnetic length is not able to boost the classification rate as expected. The result is a similar overall classification rate of 0.65.

Ch. 6 Vehicle Reidentification by Wireless Sensor Networks

Vehicle reidentification is the process of matching the detection of a vehicle at different locations. This allows tracking of vehicles from point to point along a road network. A single detector station measures traffic parameters such as speed and occupancy at a specific location. These measurements are assumed to be representative of the traffic beyond the station's location. However, this assumption may not be valid when the traffic becomes congested. The limitation of point data raised interest in vehicle reidentification, which can provide section measurements including travel time, travel time variability, section density and origin/destination demand [6.1].

The section travel time and density are useful parameters for traffic planning, control and management. The traffic parameters obtained from vehicle reidentification can be used in evaluating the performance of traffic systems and serve as inputs to advanced transportation management and information systems (ATMIS) to ease the traffic congestion [6.2].

In this chapter, current technologies for section measurements and reidentification are reviewed in section 6.1. The proposed data processing and Max-Of-Max (MOM) reidentification scheme are discussed in section 6.2. The corresponding experimental

analysis and results are presented in section 6.3.

6.1 Review of Section Measurements and

Reidentification Technologies

Common vehicle detection technologies include inductive loop, video, infrared, ultrasonic,

microwave, acoustic, magnetic, road tube and piezoelectric [Ch. 2]. Most of the time, only

point measurements such as traffic count, point speed and occupancy are generated by

these detection systems. In fact, however, more information, including the raw vehicle

signature, can be extracted from these systems for the purpose of section measurements and

vehicle reidentification.

Since point measurements are the predominant source of traffic information, one class of

techniques has been developed to estimate section travel time from these measurements.

Instead of reidentifying vehicles individually, the data are applied to a stochastic traffic-

flow model to generate a statistical estimate of the travel time. The model in [6.3] model

assumes that vehicles arriving at an upstream point during a given interval of time have a

common probability distribution of travel times to a downstream point. And the travel time

distribution is estimated by an approximate relationship between flow, occupancy and

speed. In [6.4], average travel time was estimated by using the cross-correlation analysis

of traffic flow data.

Vehicle reidentification techniques are used to generate traffic section measurements. It is accomplished by matching signatures of individual vehicles or platoons [6.5]. Common technologies include license plate matching [6.6], laser-based [6.7] or radar-based [6.8] systems that give vehicle lengths and profiles, weigh-in-motion (WIM) systems [6.9] that give axle profiles and video-based system [6.10, 6.11] that use image processing to extract lengths, profiles and colors of vehicles.

Reference	Methodology	Data	Results
[6.6] Turner S. et al.	License plate matching by video with OCR	32 hrs of video 90,000 vehicles	75% of the license plates are matched
[6.7] Cheng H. et al.	Vehicle profiles matching by laser-based system	Not available	Not available
[6.8] Urazghildiiev I.R., et al.	Vehicle height profiles matching by microwave radar	1709 vehicles	98% are correctly classified into 5 classes
[6.10] MacCarley, et al.	Video-bases vehicle signature tracking	17 minutes of video 243 vehicles	False-positive: 7.1% False-negative: 11.8%
[6.15] Ritchie, S.G., et al.	Feature extraction from inductive signatures	140 vehicles 50% for training 50% for testing	97% matching rate

Table 6.1.1 Summary of pervious researches on vehicle re-identification

Traditional inductive loop detectors are designed to focus on the vehicle detection task. New detector cards were introduced to provide inductive signatures for vehicle classification and reidentification. Since inductive loops are still the dominant surveillance system in the industry, using the existing infrastructure for vehicle reidentification is highly cost effective. Previous studies [6.12, 6.13, 6.14] show that vehicle reidentification can be done using the inductive signatures that give lengths and profiles of vehicles. However, the standard 6' square configuration of the loops induces a smoothing effect on the signature that removes distinctive features from it. A newly designed inductive loop with a long and thin geometry, called the blade sensor [6.15] partly overcomes the smoothing effect. Using the blade sensor with a high-scanning-speed detector card, detailed inductive loop signatures can be obtained with the resolution of the HMC1051Z magnetic sensor [Ch. 3]. These detailed inductive signatures, provide a higher identification rate [6.15].

There are systems that can obtain section measurements without using a fixed surveillance station [section 2.1]. These systems employ in-vehicle beacons that allow system-wide tracking of such vehicles, which include Global Positioning System (GPS) [6.16], RFID-based Automatic Vehicle Identification (AVI) [6.17], cellular telecommunication [6.18], toll tags, other in-vehicles tracking beacons and the associated infrastructure for collecting the position of the vehicles. Nevertheless, privacy concerns may prevent their large scale deployment. So vehicle reidentification systems that track vehicles anonymously are preferred.

6.2 Data Processing and Reidentification Schemes

Detailed magnetic signature examples collected from an array of sensor nodes are presented in section 6.2.1. Some of their characteristics related to reidentification are presented. The proposed Max-Of-Max (MOM) reidentification scheme used on the correlation coefficient of Average-Bar data is presented in section 6.2.2.

171

6.2.1 Magnetic Signature Array Examples

In order to get a more detailed picture of the magnetic signature of vehicles, an experiment was conducted to collect signatures from an array of 7 sensor nodes, sampling at 128Hz on each of the 3 axes. They were placed in a line perpendicular to the direction of travel, with a separation of 6" from each other. The layout of sensor nodes is shown in Fig. 6.2.1.1. Signatures of 7 test vehicles are presented and a list of the test vehicle models is shown in Table 6.2.1.1.

Fig. 6.2.1.1 The layout of 7 sensor nodes used for collecting magnetic signature array examples

	Models	Picture
(a)	Buick Le Sabre 97	
(b)	Another signature sample from the same vehicle in (a)	
(c)	Buick Le Sabre 97, a second vehicle in the same model as (a)	
(d)	Toyota Corolla 89	
(e)	LandRover Range Rover 96	
(f)	Ford Taurus 96	
(g)	Ford Taurus 2000	

(h)	Ford WindStar (Van)	

Table 6.2.1.1 List of the test vehicle models used for collecting magnetic signature array
examples

The magnetic signatures are transformed into the Average-Bar format [section 5.3.2]. The

transformed data from each of the 7 sensor nodes are combined into a single color plot,

with the horizontal axis as the sensor node index, 1 to 7, and the vertical axis as the

Average-Bar index, 1 to 20. The results are shown in Fig. 6.2.1.2 below:

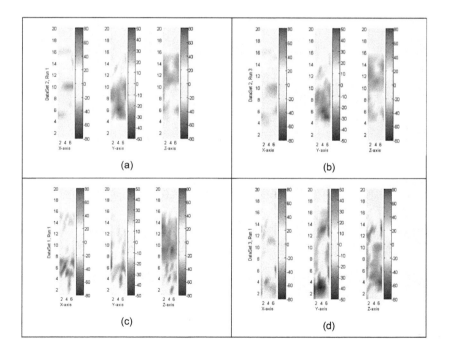

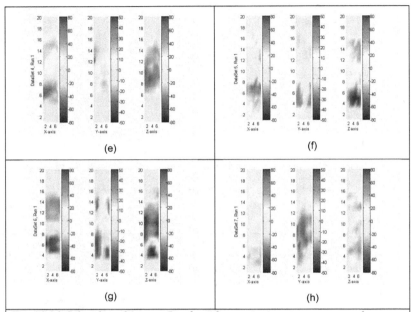

Fig. 6.2.1.2 Color plots of Average-Bar data of magnetic signature array examples

(a) and (b) are two samples collected from the same vehicle (Buick Le Sabre 97). The maximum combined correlation coefficient [section 6.2.2, 6.3] of the 3-axis signatures between them is 0.99. This high value demonstrates that the magnetic signature of a vehicle from an array of sensor nodes is reproducible.

(b) and (c) are two samples collected from two different vehicles of the same model (Buick Le Sabre 97). The maximum combined correlation coefficient of the 3-axis signatures between them is only 0.78, which is much smaller than the typical level for the *same* vehicle (0.9) [section 6.2.2, 6.3]. This demonstrates an important characteristic that vehicles of the same model do not necessarily give identical magnetic signature. The change in signatures can be explained by the differences in hardware settings and equipments in the vehicles. This implies that vehicles in the same model and color which cannot be distinguished by video and other length-based reidentification systems might be distinguished by their magnetic signatures.

These examples also re-confirm the lateral offset characteristic noted in section 5.2.2, which states that the magnetic distribution is not laterally symmetric in general. It also justifies the use of an array of sensor nodes to account for the difference in lateral offset of when vehicles go through different surveillance stations.

6.2.2 Max-Of-Max (MOM) Reidentification Scheme

Similar to the classification schemes discussion in Ch.5, the reidentification scheme proposed in this section is aimed at using the wireless sensor networks as a deployment platform. Again, in order to minimize the communication load on the sensor nodes and maximize the system lifetime, the same transformation processes described in section 5.3 are used for the reidentification system. Since the Average-Bar transformation gives the best performance in the experiments so far, this data format is used in the following analysis.

The proposed Max-Of-Max (MOM) scheme uses the Average-Bar data from all the senor nodes in the array. The maximum combined correlation coefficients of the 3-axis signatures

175

among different sensor nodes are calculated. The maximum one among different signature

sets within a certain time window will be found. If this maximum correlation coefficient is

larger than a pre-defined threshold value, a positive reidentification result is issued. The

detailed algorithms of this scheme are presented as follow:

i, The signature set (S_i) of vehicle (i) includes all the 3-axis Average-Bar transformed data

($B_{i,j}$) from the (M) sensor nodes in the array

$$B_{i,m} = \{b_{i,m,X} \quad b_{i,m,Y} \quad b_{i,m,Z}\}$$
$$S_i = \{B_{i,1} \quad B_{i,2} .. B_{i,M}\} \tag{Eq. 6.2.2.1}$$

ii, Signature sets (S) collected from station (A) at time (t) are stored in database (D_A)

$$D_A \in \{(S_{A1}, t_{A1}) .. (S_{Ai}, t_{Ai})\} \tag{Eq. 6.2.2.2}$$

iii, When a signature set (S_{Bk}) of vehicle (k) is collected from station (B) at time (t_{Bk}), a

time window (T_{ABk}) is calculated according to the travel distance between (A) and (B) with

a reasonable speed range (e.g. 5 to 80 mph). This represents the time window of vehicles

detected at station A that can reach station B at time (t_{Bk})

$$T_{ABk} = [t_{ABk0}, t_{ABk1}] \tag{Eq. 6.2.2.3}$$

iv, Signature sets are extracted from database (D_A) according to the time window (T_{ABk})

$$D_A(T_{ABk}) \in \{(S_{Ai}, t_{Ai}) \quad \forall i \text{ in } D_A : t_{ABk0} \leqslant t_{Ai} \leqslant t_{ABk1}\}$$

$$\tag{Eq. 6.2.2.4}$$

v, The maximum combined correlation coefficient (C_{max}) between (S_{Bk}) and $D_A(T_{ABk})$ is

calculated by comparing the products of correlation coefficients of each of the 3-axis data

between (S_{Bk}) and each of the signature sets in $D_A(T_{ABk})$. corr(b1, b2) is a function that will

return the correlation coefficient of two vectors (b1, b2)

176

$$CORR\left(B_{i,j}, B_{k,m}\right) = \left| corr\left(b_{i,j,X}, b_{k,m,X}\right) \right| \times \left| corr\left(b_{i,j,Y}, b_{k,m,Y}\right) \right| \times \left| corr\left(b_{i,j,Z}, b_{k,m,Z}\right) \right|$$

$$CORR\left(S_{Bk}, S_{Ai}\right) = \left\{ CORR\left(B_{Bk,m}, B_{Ai,j}\right) \quad \forall m : 1 \leqslant m \leqslant M \quad \forall j : 1 \leqslant j \leqslant M \right\}$$

$$CORR\left(S_{Bk}, D_A\left(T_{ABk}\right)\right) = \left\{ CORR\left(S_{Bk}, S_{Ai}\right) \quad \forall i \ in \ D_A\left(T_{ABk}\right) \right\}$$

$$C_{max}\left(S_{Bk}, D_A\left(T_{ABk}\right)\right) = \max\left(CORR\left(S_{Bk}, D_A\left(T_{ABk}\right)\right) \right)$$

(Eq. 6.2.2.5)

vi, If (C_{max}) is higher than a pre-defined threshold value ($C_{threshold}$), the vehicle (k) is declared to be reidentified by the system as vehicle (i) in database (D_A), and not reidentified otherwise

$$REID\left(S_{Bk}\right) = \begin{cases} maxarg_i\left(\left\{ \max\left(CORR\left(S_{Bk}, S_{Ai}\right)\right) \quad \forall i \ in \ D_A\left(T_{ABk}\right)\right\}\right) & if\left(C_{max} \geqslant C_{threshold}\right) \\ Null & Otherwise \end{cases}$$

(Eq. 6.2.2.6)

vii, If more than one vehicle give a maximum combined coefficient that is very close to (C_{max}), and higher than ($C_{threshold}$), the one with a larger sum over the combined coefficients in its signature set will be taken.

$$REID\left(S_{Bk}\right) = maxarg_i\left(\left\{ sum\left(CORR\left(S_{Bk}, S_{Ai}\right)\right) \quad \forall i \ in \ D_A\left(T_{ABk}\right) : \left| \max\left(CORR\left(S_{Bk}, S_{Ai}\right)\right) - C_{max}\right| \leqslant \varepsilon\right\}\right)$$

(Eq. 6.2.2.7)

A drawback of this scheme is that the generation of correlation coefficients is computationally expensive, with processing time on the order of (N^2). This is another reason for using the Average-Bar data, which is helpful in reducing the computational burden. An experimental evaluation of this reidentification scheme is presented in the next section.

6.3 Experimental Analysis and Results

In section 6.3.1, a preliminary reidentification experiment (dataset R1) that involves 7 test vehicles is presented. In section 6.3.2, the analysis and results of a reidentification experiment (dataset R2) on left-turning vehicles are presented.

6.3.1 Preliminary Reidentification of 7 Test Vehicles (dataset R1)

This experiment is aimed at testing the feasibility of using Average-Bar data [section 5.3.2] of the magnetic signatures with the proposed MOM reidentification scheme [section 6.2.2]. The signature examples presented in section 6.2.1 were also generated from this experiment conducted in Richmond Field Station at July 12th, 2005. An array of 7 sensor nodes was placed in a line perpendicular to the direction of travel, with a separation of 6" from each other. A photo of the experimental setup for the nodes is shown in Fig. 6.2.1.1. A list of the test vehicle models is shown in Table 6.2.1.1.

Each test vehicle was driven over the sensor nodes 5 times, with a different lateral offset intentionally added by the driver. The transformed data from each of the 7 sensor nodes are combined into a single color plot, with the horizontal axis as the sensor node index from 1 to 7 and the vertical axis as the Average-Bar index from 1 to 20. The color plots of all 5 runs of test vehicle (g) (Ford Taurus 2000) are shown in Fig. 6.3.1.1 below. Although the signatures from individual node may look different in different runs because of the change in lateral offset; the overall signatures shown in the color plots are similar. Among the data from test vehicle (g), an average maximum combined correlation coefficient of 0.93 is obtained.

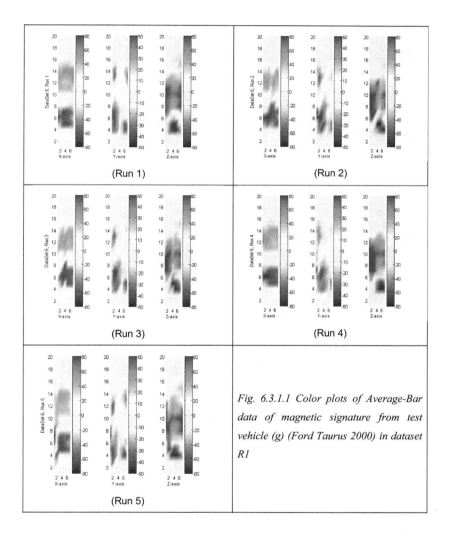

Fig. 6.3.1.1 Color plots of Average-Bar data of magnetic signature from test vehicle (g) (Ford Taurus 2000) in dataset R1

Using the 35 collected signatures, a reidentification analysis was simulated by taking one signature at a time as the target test sample, and comparing it against a dataset of 7 signatures with one from each test vehicle, with ($C_{threshold}$) = 0.9. All possible combinations

179

within the database were tested with the reidentification scheme described in section 6.2.2.

A summary of the results is shown in Table 6.3.1.1.

Target Vehicle	Correct Reidentification Rate [%]			
	corr(X)	corr(Y)	corr(Z)	corr(X*Y*Z)
(a)	100.0	71.2	100.0	100.0
(b)	100.0	91.8	94.1	100.0
(c)	82.8	79.3	79.3	93.0
(d)	100.0	83.2	98.0	100.0
(e)	100.0	100.0	100.0	100.0
(f)	100.0	100.0	100.0	100.0
(g)	79.2	98.2	70.1	99.0
Avg	94.6	89.1	91.6	98.9

Table 6.3.1.1 Summary of reidentification results in dataset R1

An average reidentification rate of 98.9% is achieved with a database of these 7 test

vehicles when all the 3-axis data are used. Although the sample size is small, this

promising result demonstrates the feasibility of using the Average-Bar data [section 5.3.2]

with the proposed scheme [section 6.2.2] on wireless sensor networks for reidentification.

6.3.2 Reidentification of Left-Turning Vehicles (dataset R2)

The information about traffic flow between origin/destination (O/D) pairs is essential for

efficient traffic planning and control [6.19]. Reidentification systems can be used for

estimating the O/D matrix. However, because of high installation and maintenance costs,

they are seldom used for urban traffic. This experiment is aimed at demonstrating how the wireless sensor networks can be used to reidentify left-turning vehicles at a local intersection, so that the turn ratio can be derived and used in O/D matrix estimation [6.20].

This experiment was also conducted in Richmond Field Station (RFS) on March 9th, 2006. Two arrays of 7 sensor nodes, sampling at 128Hz on each of the 3 axes, were placed in a line perpendicular to the direction of travel. One was located at the end of a straight section (station A); another one was located at the beginning of a straight section after a left turn from station A (station B). Photos of the experimental setup are shown in Fig. 6.3.2.1. The sensor nodes were placed in such a way that no right-turning vehicles would drive over the nodes at station B. Only straight-passing and left-turning vehicles are considered in the analysis.

Fig. 6.3.2.1 Experimental setup for the left-turning reidentification experiment (R2)

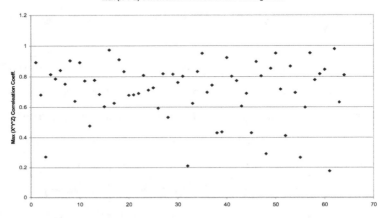

Max (X*Y*Z) Correleation Coeff. for all Left Turning cases

Fig. 6.3.2.2 Maximum combined correlation coefficient of all the left turning cases in R2

Ground truth of the experiment was obtained by manual observation. A total of 80 vehicles were recorded at the test site in 4 hours. 10-Average-Bar transformation was applied on these data. A plot of the maximum combined correlation coefficient (C_{max}) of all the left-turning cases is shown in Fig. 6.3.2.2. According to the distribution of the (C_{max}), a value of 0.56 was used for ($C_{threshold}$) in the reidentification scheme [section 6.2.2]. This threshold value is relatively small compared to 0.9 used in experiment R1. It is justified by the fact that some difference in the magnetic signatures is induced by the change in travel direction [section 5.2.1] after turning left. Since the traffic flow at the test site is low (20 veh/hr), a "time window" of last 5 vehicles at station (A) is used for (T_{ABk}). The reidentification results are summarized in Table 6.3.2.1.

	Observed	Correctly reidentified			Correct %
Left-Turning	64	46			71.9
			False Positive	7	
			False Negative	11	
Straight-Passing	16	12			75.0
Total	80	58			72.5

Table 6.3.2.1 Summary of reidentification results in dataset R2

An overall reidentification rate of 72.5% is achieved in this test. Again, although a sample size of 80 is small, this positive result demonstrates the potential of large scale deployment of reidentification systems on freeway and urban area by wireless sensor networks.

Ch. 7 Evolution of Intelligent Transportation System

The proposed wireless sensor networks have the potential to revolutionize the whole traffic surveillance and control industry into one that is scalable and deployable everywhere in traffic networks. Its feasibility for large scale deployment can impact the associated Intelligent Transportation System (ITS) in many applications. In this chapter, the impacts of using such a surveillance system on a large scale are provided in section 7.1. And the possibility of extending the sensor networks into a multi-function wireless surveillance system is studied in section 7.2.

7.1 Impact of Large Scale Deployment

In 2005 Urban Mobility Report [1.1], the benefits from the implementation of four ITS technologies are studied: traffic signal coordination, arterial street access management, freeway entrance ramp metering and freeway incident management. The benefits are estimated to be 336 million hours of delay reduction and $5.6 billion in congestion savings for the 85 urban areas in 2003. If these technologies were deployed on all the major roads, an estimated 613 million hours of delay and more than $10.2 billion would be saved.

However, the large scale deployment of ITS technologies is discouraged by the high life-cycle cost [section 2.1.3] and large traffic delay caused by the installation of inductive loop detectors. The proposed wireless sensor networks provide the same functionalities as inductive loop systems, but at a much lower cost and delay. Its wireless capability and configuration flexibility [section 2.2] give extra advantages to the large scale deployment of many ITS systems. Examples of traffic signal control, on-ramp metering and parking management are provided in this section.

7.1.1 Traffic Signal Control

The objective of a signal control system is to minimize the delay experienced by vehicles traveling through a network of intersections by manipulating the traffic signal plans. A summary of recent advancements in traffic signal control is provided in [7.1, 7.2]. Using ITS technologies, new strategies have been developed to address dilemma zone [7.2] problems of signalized intersections. These strategies can be described as follows:

i, Detection Control System (DCS) [7.3] monitors approaching traffic of an intersection and minimizes the number of vehicles that will fall into the dilemma zone, by controlling the end time of the green phase. Besides speed estimate, classification is needed to identify approaching large trucks and apply the corresponding controls.

ii, Platoon Identification and Accommodation (PIA) system [7.4] estimates the length and time of arrival of approaching platoons. The traffic signals are optimized to allow them to pass the intersection smoothly.

iii, Advanced Warning of End of Green System (AWEGS) [7.5] provides advance warning for end of green to drivers by flashing beacons located about 700 feet upstream of the intersection. A picture of such a beacon is shown in Fig. 7.1.1.1.

Fig. 7.1.1.1 Picture of a flashing beacon for end of green warning [7.6]

Advance vehicle detection is the most critical component for the implementation of all these signal control strategies. For a high-speed approach, advanced detection is required at 800 to 1000 feet upstream of the intersection, as well as the speed and classification of vehicles at real-time. If inductive loop detectors are used, the cost of setting up a communication link between the traffic signal controller and the upstream detection station is estimated to be $15,000 [7.6] for a two lanes approaches. Such an expensive data transmission cost significantly reduces the benefit-to-cost ratio of these systems.

On the other hand, the proposed wireless sensor networks virtually add no extra cost for the data transmission as wireless communication capability is already built-in. Taking advantage of the relatively low life-cycle cost [section 2.1.2] of this surveillance system, the signal control systems can be cost-effective enough to be deployed at all the intersections of major arterials to maximize the traffic network capacity. A typical configuration of such a system is shown in Fig. 7.1.1.2 [7.7].

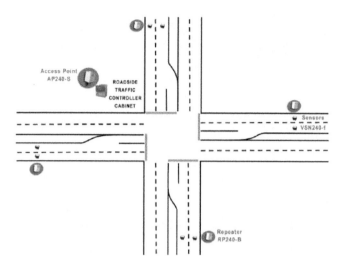

Fig. 7.1.1.2 Typical configuration of wireless sensor networks (VSN240) for advance detection at an intersection [7.7]

7.1.2 On-Ramp Metering

On-ramp meters regulate the flow of traffic on freeway entrance ramps using traffic signals. They are designed to create more space between entering vehicles so those vehicles do not disrupt the main stream traffic flow. The number of entering vehicles for short distance trips is also reduced, as the drivers are encouraged to use the local streets to avoid the ramp wait time [7.8]. It has been reported that on-ramp metering was able to reduce delay by 102 million person-hours in 2003, approximately 5% of the congestion delay on freeways where ramp metering was in effect [1.1]. Because of its high effectiveness, it has been widely deploy and recommended to the Federal Highway Administration as an essential tool to address the congestion problem [7.9].

187

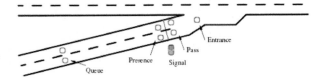

Fig. 7.1.2.1 Typical configuration of an on-ramp metering system with inductive loop detectors [7.10]

Whenever the on-ramp demand exceeds the metering rate, a queue will form in front of the signal. The on-ramp storage capacity is usually limited. If the queen length exceeds this capacity, it will spill into connected streets and interfere with the local traffic. A typical configuration of on-ramp metering system with inductive loop detectors is shown in Fig. 7.1.2.1. The current practice to regulate the queue length is to adjust the metering rate if detections are trigged from the queue detectors. However, it was found that this scheme leads to oscillatory behavior and under-utilization of on-ramp storage capacities [7.11, 7.12].

The proposed wireless sensor networks can be used instead of the inductive loops to bring down the transmission cost to virtually zero [section 7.1.1]. Multiple sensor nodes can be placed along the on-ramp's storage lanes to provide multiple detection zones for the system. This flexible configuration allows the queue length to be monitored at the desirable resolutions of the control strategies. The on-ramp metering system can also communicate with the surveillance system on the freeway and the traffic light control system near by over the same wireless networks. Such a large scale deployment framework allows global optimized control strategies [7.13] to be designed and implemented.

7.1.3 Parking Guidance and Information System (PGIS)

Parking is a universal problem in most metropolitan areas that already suffer from heavy

traffic congestion and air quality degradation. Limited parking space and the lack of

information on parking availability make the parking search time unreasonably long. This

undesirable parking search traffic leads to additional congestion, air pollution and driver

frustration. Increasing parking space is discouraged by the limited land space and its high

cost in urban area. Therefore, Parking Guidance and Information System (PGIS) [7.14] is

introduced to minimize the parking search traffic.

PGIS has been implemented in several European and Japanese [7.14] cities since 1970s.

Early systems provide parking availability information on Variable Message Signs (VMS)

by counting the number of vehicles entering and leaving the facilities with inductive loops.

Recent systems provide real-time information of number of opening by monitoring the

occupied spaces. Available sensor technologies include inductive loops, video image

processing, ultrasonic, infrared and microwave [7.15, section 2.1]. Besides VMS, the

information is delivered via mobile phones, radio, internet and in-vehicle navigation

systems.

PGIS is usually deployed at transit stations to encourage the use of public transits [7.16]. A

significant relationship between transit usage and parking spaces was suggested by research

[7.17]. The system can also be used to the deliver additional information like transit

departure time and traffic incidents. PGIS is used to provide exact location of open parking slots in large parking facilities (e.g. airport) [7.18]. These parking management applications require surveillance systems with a large coverage of detection zones. Video Image Processing (VIP) systems [section 2.1] do fit the requirements except that it is greatly affected by the environmental and lighting conditions. They are subjected to changing environment in out-door parking lots and poor lighting condition in in-door parking facilities.

On the other hand, wireless sensor networks can fit into any kind of parking facilities with sensor nodes that can be placed anywhere. The simple installation structure and procedure makes the implementation scalable, which is an important issue when thousands of sensor nodes are used. Its wireless communication capability also allows it to provide detailed parking guidance to in-vehicle navigation system in the future. Because of a relatively low-activity of parking detection is needed, the system life cycle can be dramatically increased to 30-40 years and decreases the life-cycle cost [section 2.1.3] accordingly. A typical configuration of wireless sensor networks for parking monitoring is shown in Fig. 7.1.3.1 [7.19].

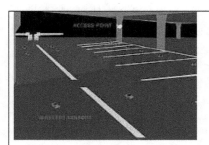

Fig. 7.1.3.1 Typical configuration of wireless sensor networks (VDS240) for parking monitoring [7.19]

7.1.4 Work Zone Management

Between 1980 and 1998, vehicle miles traveled increased 72% while the number of lane miles increased only 1% [7.20]. In order to match this traffic growth, there is an urgent need for the construction and rehabilitation of roadways. However, these work zones cause congestion and safety problems. It is reported that more than 800 fatalities and 37,000 injuries are caused by the work zones in U.S. every year [7.21]. Therefore, ITS technologies are applied in large scale work zones to let vehicles travel through and around the zones more safely and efficiently.

Traffic surveillance systems are used to monitor traffic in real-time. The traffic condition and lanes closure information are delivered to the public via VMS, mobile phones and internet. Travelers can adjust their routes to avoid congestion or delay. Safety is also enhanced by providing advance notices of slowed or stopped traffic. The use of ITS in work zones is proven to be successful [7.21] and has been included in the National ITS Program Plan.

Video Image Processing (VIP) systems are usually used to monitor the traffic around the work zones as it provides video images for manual observation and incidents detection. However, the use of VIP systems brings the capital cost to a high level. In a case studied in [7.21], $2.4 million was spent to setup VIP systems at a work zone in Lansing, Michigan. Considering the life-cycle cost of the proposed wireless sensor networks is roughly half of that of a VIP system, $1.2 million could be saved if the sensor networks were used [7.21]. Typical work zones seldom last longer than 5 years, which is only half of the sensor networks' life-time. This implies the sensor nodes can be reused at another work zone after recharging. This huge cost reduction in the surveillance system makes the work zones management cost effective enough to be deployed at all work zones no matter what size the construction project is.

7.2 Multi-Functions Wireless Surveillance System

Multi-functions wireless surveillance systems can be developed by adding other sensing modalities to the traffic surveillance systems. An important one is the modality for sensing road conditions which is presented in section 7.2.1. The wireless communication capability of the sensor networks also allows it to talk to other ITS systems. Since the sensor nodes are located on the pavement, the networks can be a very useful tool in the Vehicle-Infrastructure Integration (VII) framework. It can be used to exchange information between different systems and extend the vehicle-infrastructure communication range. Its applications in VII are presented in section 7.2.2.

7.2.1 Road Conditions Sensing Modality

There were over 1.4 million road conditions-related crashes in 2001, causing over 615,000 injured, over 6,900 dead and over 1 billion hours delay [7.25, 7.26]. The monitoring of adverse weather and road conditions in real-time is essential for safety enhancement and roadway maintenance [7.22].

The detection of water, ice, snow, fog and lighting conditions can be used to determine whether the visibility, weather and road conditions are safe for driving. Warnings of potential hazard (e.g. icing on bridges) can be given to drivers via VMS, in-vehicle navigation system or other media. This safety enhancement is especially useful when the conditions are not obvious to the drivers (e.g. poor lighting conditions, black ice). Traffic signal controls [section 7.1.1] can be adjusted according to the change in road conditions. This information is also used in roadway maintenance to determine the type and extent of surface treatments needed for keeping the roadway drivable.

According to the experience from Minnesota [7.23], such systems improve traffic safety, enlarge roads capacity and allow an effective planning of future investment. National surface transportation weather observing and forecasting system (Clarus) [7.25] has been initialized by the Federal Highway Administration (FHWA) to implement road conditions monitoring system nationwide.

State-of-the-art sensor technologies [3.4] are ready for these applications. For example, a micro-sensors board designed for weather monitoring is developed under the TinyOS project [section 3.2], known as the MICA weather board MTS400 [7.23]. It includes sensors for measuring temperature (ice and snow), humidity (rain and fog), light intensity (lighting condition) and air pressure. In addition, acoustic sensors can be used to estimate wind speed and accelerometers can be used to estimate weight of vehicles. All of these sensors are integrated into a board with the size of a name card. A picture of MTS400 is shown in Fig. 7.2.1.1 [7.24]. This sensor board is designed to be used with MICA sensor nodes family [section 3.2.2], and it can be integrated into Sensys' sensor nodes [section 3.2.5] easily.

Fig. 7.2.1.1 Picture of MICA weather board MTS400 manufactured by Crossbow [7.24]

The power consumption of this additional sensing modality is relative small as the sampling frequency can be as low as once every 15 minutes. Its effect on the system lifetime can be further minimized by using a number of sensor nodes at the same site to take care of the tasks alternatively, or by using an extra node to concentrate on monitoring the road conditions. This plug-and-play feature allows additional sensing modalities to be added to the wireless sensor networks surveillance system without modifying the system framework. Therefore, deployments of different sensing capabilities can be customized according to specific applications and locations.

7.2.2 Vehicle-Infrastructure Integration (VII)

Half of the 43,000 deaths that occur each year in U.S. result from vehicles entering or leaving the freeway, and traveling unsafely through intersections [7.27]. Aimed at minimizing these accidents, and the associated traffic delay and cost, the U.S. Department of Transportation (USDOT) has proposed the VII initiative. It studies the development of a nationwide wireless communication infrastructure that allows vehicle-vehicle and vehicle-infrastructure communications. With the push from the government, it is foreseen that

194

every car manufactured in US will be equipped with a standardized wireless communication device and a GPS unit for integrating with the VII infrastructures within 10 years. The Dedicated Short Range Communications (DSRC) protocol is adopted as the standard. Radio spectrum at 5.9 GHz is specifically allocated for the use of DSRC [7.28].

Vehicles with communication capability can serve as anonymous probe vehicles for collecting traffic information and road conditions [section 2.1.1.3]. The data can be fused with the existing surveillance database and enhances the real-time traffic control [section 7.1]. Traffic accidents are prevented by warning the drivers when their vehicles are entering an intersection unsafely or when they are running dangerously close to other vehicles. A good example of such applications is Intersection Decision Support [7.29], which activates left-turn warning signs if other vehicles are approaching the intersection from the opposite direction, so that Left Turn Across Path/Opposite Direction (LTAP/OD) and Left Turn Across Path/Lateral Direction (LTAP/LD) crashes can be avoided [7.29]. A typical system configuration of IDS is shown in Fig. 7.2.2.1 [7.29].

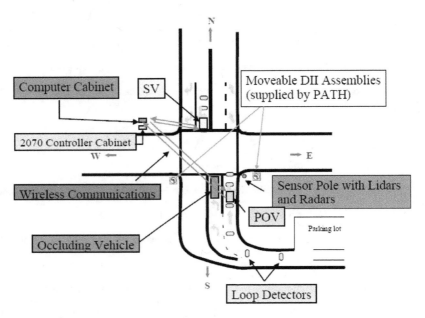

Fig. 7.2.2.1 Typical system configuration of Intersection Decision Support (IDS) [7.29]

The proposed wireless sensor networks can be used with VII in two approaches:

i, Once VII is deployed in a nationwide sense, the road will be filled with a mixture of

vehicles and infrastructures with and without communication capability. The sensor

networks can provide surveillance support to the system for vehicles and infrastructures

that do not have the communication capability. This allows the associated ITS applications

to be used independent of the VII penetration rate.

ii, The sensor networks can also adopt the standardized DSRC protocol so that they can be

used to extend the range of vehicle-vehicle and vehicle-infrastructure communications, by

acting as repeater nodes. This mixed system lowers down the deployment cost by minimizing the need for building new infrastructure along the side of road. This makes the systems cost efficient enough to be deployed in large scale, especially for applications that need advance detection [section 7.1.1].

Ch. 8 Conclusion

Between 1980 and 1998, vehicle miles traveled increased 72% while the number of lane miles increased only 1% [7.20]. The current pace of improvement in transportation system is not sufficient to keep up with even a slow growth in the traffic demands in most major urban areas. There is a great need for advanced surveillance capabilities to complement the rapid deployment of ITS strategies. Because of the highly intrusive characteristic of inductive loop detectors, the quest for researching a reliable and cost-effective alternative system, which can provide traffic data at the same accuracy level as inductive loop systems, has been underway for some time.

In this dissertation, details of design and development of a novel wireless sensor networks based traffic surveillance system are presented, which has a detection accuracy as good as that of inductive loop detectors [section 4.4]. It offers a very attractive alternative to inductive loops for traffic surveillance. The sensor networks have a much higher configuration flexibility, which allows the system to be scalable and deployable everywhere in the traffic networks. The availability of these data opens up new opportunities for intelligent traffic operations and control [Ch. 7]. Having a lower system life-cycle cost than inductive loop, video and radar detector systems [section 2.1.2], the sensor networks are cost-effective enough for large scale deployment.

Total coverage of surveillance does not need to stay in the simulation stage anymore. The proposed wireless sensor networks have the potential to revolutionize the whole traffic surveillance and control industry [Ch. 7]. A summary of contributions from this research project is presented in section 8.1. And several potential future developments of this system are provided in section 8.2.

8.1 Summary of Contributions

For maximizing the benefits from all the ITS technologies, a large scale deployment of traffic controls on all major freeways and local streets would be an essential step to be taken [1.1]. Therefore, real-time traffic information at all these sites is required. In this research project, wireless sensor networks were developed and tested as a traffic surveillance system with detection accuracy as good as that of inductive loop detectors [section 4.4]. With such a flexible and cost-effective surveillance system [section 2.2], large scale deployment of ITS technologies can be achieved.

The first generation of sensor node prototype specially designed for this project was named TrafficDot [section 3.2]. Its basic components include two magnetic sensors (HMC1051Z [3.14]), a microcontroller with integrated radio (MICA2DOT [3.15]), a battery (TL-5135 [3.16]) and a protective cover (SmartStud [3.17]) made by high impact polycarbonate and designed to be placed on pavement. Even though this very first prototype does not have complete functionality, it shows us a very promising future in the application of wireless sensor networks on traffic surveillance. It also provided important directions for improving the design and development of second and third generation prototypes, which eventually became commercialized [3.18].

Traffic surveillance starts with detecting the presence of a vehicle. This very first stage in the surveillance system determines the final performance of all the dependent applications. Signals from magnetic and acoustic sensors are analyzed for designing detection algorithm [section 4.1, 4.2]. An efficient and robust real-time detection algorithm for these sensors, called Adaptive Threshold Detection Algorithm (ATDA) [4.1] was developed. A correct detection rate higher than 98% (777 / 791) was achieved in a real-time test (D5) [section 4.4.5] conducted at a local intersection, while the inductive loops detectors gave a 14% over-counts in the same test.

Vehicle classification is an important source of information for transportation design and management. However, conventional classification approaches either involve extensive computation or require a high volume of communication flow between the sensor nodes and base station, which are undesirable for the implementation on wireless sensor networks with limited computation resources and high power consumption in communication.

Vehicle magnetic signatures were collected and analyzed for designing new classification schemes aimed at a platform with very limited computation resources [section 5.3]. Novel transformation processes, Average-Bars and Hill-Patterns [section 5.3.2], are proposed to transform the vehicle signature into a highly compressed format with fixed size. After principal component analysis, the data is classified by k-Nearest Neighbor (k-NN), Support

200

Vector Machine (SVM) and direct hill-pattern matching [section 5.3.4]. A classification rate of 85% was achieved with a dataset of 864 trucks with the use of k-NN classifier on Average-Bars data for classifying FHWA class {5, 6} (3-or-less-axle) vs {8, 9, 11} (4-or-more-axle). The overall classification rate was 65% for classifying FHWA class {5 vs 6 vs 8 vs 9 vs 11} [section 5.4.5]. This simple classification scheme can be used to classify passenger vehicles from trucks, and 3-or-less-axle trucks from 4-or-more-axle trucks.

The limitation of point data raised interest in vehicle reidentification, which provides section measurements including travel time, travel time variability, section density and origin/destination demand [section 6.1]. These data are essential parameters for effective traffic planning, control and managements. A Max-Of-Max (MOM) reidentification scheme using the Average-Bas data is proposed [section 6.2]. A reidentification rate of 72.5% (58 / 80) was achieved in a left-turning reidentification test (dataset R2) [section 6.3]. This demonstrated that the reidentification system is portable and deployable for different applications.

Large scale deployment of ITS technologies is discouraged by the high life-cycle cost [section 2.1.3] and huge traffic delay caused by the installation of inductive loop detectors. The proposed wireless sensor networks provide the same functionality as inductive loop systems, but at a much lower cost and delay. Its wireless capability and configuration flexibility [section 2.2] give extra advantages to the large scale deployment of many ITS systems, such as traffic signal control, on-ramp metering and parking management [section 7.1].

8.2 Future Developments

The proposed wireless sensor networks are extremely flexible platforms that can be fitted into all kinds of ITS applications. For each of these applications, there exist many different extensions of the system that can enhance the performance or provide new functionality. Four potential future developments are suggested below: energy harvesting, installation-in-motion, multi-function networks and real-time implementations.

Energy Harvesting

The lifetime of Wireless Sensor Networks (WSN) directly depends on its power source. In fact, the energy constraint is a dominant factor of system design trade-offs for small embedded sensor devices. The scaling down in size and cost of microcontroller and sensor has outpaced that of a battery. This makes the power source account for a growing proportion of the cost of WSN. This cost is further magnified by the maintenance cost for either replacing or recharging the batteries on a regular basis.

Harvesting Technology	Power Density
Solar cells (outdoors with sunlight)	$15 \text{ mW} / \text{cm}^2$
Piezoelectric (shoe inserts)	$330 \text{ }\mu\text{W} / \text{cm}^2$
Vibration (microwave oven)	$116 \text{ }\mu\text{W} / \text{cm}^2$
Thermoelectric ($10°C$ gradient)	$40 \text{ }\mu\text{W} / \text{cm}^2$
Acoustic (100 dB)	$960 \text{ nW} / \text{cm}^2$

Table 8.2.1 Power densities of harvesting technologies [8.1]

Fortunately, a promising technique to forestall this energy problem is emerging: energy harvesting from the environment. Making use of the energy sources ubiquitous to the operating space of the WSN raises the possibility of unlimited lifetime. Table 8.2.1 shows the power density of different harvesting technologies [8.1]. Among the wide variety of harvesting modalities, solar energy harvesting through photo-voltaic conversion provides the highest power density - 15 mW/cm^2 in sunlight. However, the design of such a module involves complex interactions of several factors, such as energy availability, solar cells characteristics, storage chemistry and capacity, and power supply requirements of specific applications. For maximizing the energy efficiency of WSN with energy harvesting, battery aware system design [8.2] and harvesting aware power management [8.3] are introduced. Deploying a wireless surveillance system with energy harvesting modalities can further extend its lifetime and reduce its life-cycle cost [section 2.1.2]. This makes the system even more attractive among all traffic surveillance technologies.

Installation-In-Motion

Even using the surface mounting type sensor nodes [section 3.2.5], an unavoidable short pavement-installation time (about 10 min) will still be needed, which involves the blocking of associated traffic. This short pavement-installation time is acceptable by allocating the work after mid-night when the traffic load is light. Nevertheless, this minimum level of traffic disruption is still a big disadvantage of the system, especially for planning a large scale deployment. Therefore, installation-in-motion approach is a potential solution that worth to spend more resources to research on. For example, a customized truck can be designed and built to automate the pavement-installation processes while it is traveling. A

GPS system can be used to identify pre-defined locations for the installation of sensor nodes. Specially designed robotic system can be used to carry out the mechanical mounting processes in a short time such that the whole installation can be completed while the truck is traveling at normal speed, inducing no traffic disruption at all. This installation-in-motion approach also makes the deployment more systematic and scalable. This is a big advantage for the system, especially for planning a nationwide deployment.

Multi-Functions Networks

Multi-functions wireless surveillance networks can be developed by adding other sensing modalities to the traffic surveillance systems. An important one is the modality for sensing road conditions [section 7.2.1]. The plug-and-play feature of sensor networks allows sensing modalities to be added to the surveillance system without modifying its framework. Therefore, deployments of different sensing capabilities can be customized according to specific applications and locations. The wireless communication capability of the sensor networks also allows it to talk to other ITS systems. Since the sensor nodes are located on the pavement, the networks can be a very useful tool in the Vehicle-Infrastructure Integration (VII) framework. It can be used to exchange information between different systems and extend the vehicle-infrastructure communication range [section 7.2.2].

Real-Time Implementations

The vehicle classification [Ch.5] and reidentification [Ch.6] algorithms are designed to be implemented on the wireless sensor networks which have a limited computation power and high communication power cost. However, all these algorithms have not been implemented on the sensor nodes yet, the processing being handled offline by Matlab so far. It is time to upgrade Sensys' VDS240 sensor nodes family to include real-time classification and reidentification modalities. This requires programming work on the sensor nodes' firmware, as well as the user-end control software. Once a full system of the wireless sensor networks is commercialized with all the detection, classification and reidentification modalities, it will be a very competitive product in the traffic surveillance industry. The market will no longer be dominated by inductive loop detectors.

Eventually, massive traffic surveillance networks will be setup for the deployments of ITS technologies on all the freeways, arteries and local intersections. Countless lives, time and money will be saved by advanced traffic controls [Ch.7]. Another page in the history of transportation system will be written, with the title "New Era of ITS Deployment". Keeping this promising future in mind, there is no doubt that more resources should be invested in the research and development of traffic surveillance systems and its associated ITS applications.

Ch. 9 References

[1.1]   David Schrank, Tim Lomax, 2005 Urban Mobility Report, Texas Transportation
Institute, The Texas A&M University System, 2005, http://mobility.tamu.edu/ums/report/

[1.2]   X. Zhang, Y. Wang and N.L. Nihan. Monitoring a freeway network in real-time
using single-loop detectors: System design and implementation. 83rd TRB Annual
Meeting, January 2004, Washington, D.C.

[1.3]   S. Oh, S.G. Ritchie and C. Oh. Real time traffic measurement from single loop
inductive signatures. 81st TRB Annual Meeting, January 2002, Washington, D.C.

[2.1]   Gordon, R.L., R.A. Reiss, H. Haenel, E.R. Case, R.L. French, A. Mohaddes, and R.
Wolcott; Traffic Control Systems Handbook, FHWA-SA-95-032, Federal Highway
Administration, U.S. Department of Transportation, Washington, D.C., Feb. 1996

[2.2]   Martin, PT; Feng, Y; Wang, X; Detector Technology Evaluation, Department of
Civil and Environmental Engineering, MPC Report No. 03-154, 2003

[2.3]   Seri Oh, Stephen G. Ritchie, and Cheol Oh. Real Time Traffic Measurement From
Single Loop Inductive Signatures, presented for 81st Annual Meeting of the Transportation
Research board, Washington D.C., January 2002.

[2.4]   Anna Pushkar, Fred L. Hall, and Jorge A. Acha-daza. Estimation of Speeds from
Single -Loop Freeway Flow and Occupancy Data Using Cusp Catastrophe Theory Model,
Transportation Research Record 1457, p149-157, Washington D.C., 1994.

[2.5]   Chao Chen, Jaimyoung Kwon, John Rice, Alexander Skabardonis and Pravin

Varaiya. Detecting Errors and Imputing Missing Data for Single Loop Surveillance

Systems, Presented in the 82nd Transportation Research Board Annual Meeting,

Washington D.C., January 2003.

[2.6]   Luz Elena Y. Mimbela and Lawrence A. Klein. A Summary of Vehicle Detection

and Surveillance Technologies used in Intelligent Transportation Systems, the Vehicle

Detector Clearinghouse, New Mexico State University, Fall 2000.

[2.7]   Craig J. Wynant; Quartz Technology for Weigh-in-Motion Sensors, Kistler

Instruments Corporation, 2002.

[2.8]   DP 121 Weigh-in-Motion Technology, Oak Ridge National Laboratory,

http://www.ornl.gov/dp121, accessed by February 11, 2003.

[2.9]   Ahmad Safaai-Jazi, Siamak A. Ardekani and Majid Mehdikhani. A Low-Cost Fiber

Optic Weigh-in-Motion Sensor, SHRP-ID/UFR-90-002, Strategic Highway Research

Program, National Research Council, Washington, D.C., 1990.

[2.10]   Wikipedia, The Free Encyclopedia, http://en.wikipedia.org/wiki/Radar

[2.11]   Duzdar, A.; Kompa, G.;Applications using a low-cost baseband pulsed microwave

radar sensor, Instrumentation and Measurement Technology Conference, 2001. IMTC

2001. Proceedings of the 18th IEEE Volume 1, 21-23 May 2001 Page(s):239 - 243 vol.1

[2.12]   ASIM Technologies Ltd., http://www.asim.ch/

[2.13]   Siemens Energy and Automation, http://automation.usa.siemens.com/index.asp

[2.14]   Kranig J., E. Minge, and C. Jones, Field Test of Monitoring of Urban Vehicle

Operations Using Non-Intrusive Technologies, FHWA-PL-97-018, Federal Highway

Administration, U.S. Department of Transportation, Washington, D.C., May 1997.

[2.15] Klein, L. A., Data Requirements and Sensor Technologies for ITS, Norwood, MA, Artech House, 2001.

[2.16] MacCarley C.A., S. Hockaday, D. Need, S. Taff, Evaluation of Video Image Processing Systems for Traffic Detection, Transportation Research Record No. 1360, National Research Council, Washington D.C., 1992.

[2.17] Mantri, S.; Bullock, D.; A neural network based vehicle detection and tracking system, System Theory, 1995., Proceedings of the Twenty-Seventh Southeastern Symposium on 12-14 March 1995 Page(s):279 - 283

[2.18] Michalopoulos, P.G., R.D. Jacobson, C.A. Anderson, and J.C. Barbaresso, Integration of Machine Vision and Adaptive Control in the Fast-Trac IVHS Program, 72nd Annual Meeting, Transportation Research Board, Washington, D.C., Jan. 1993.

[2.19] NOVAX Industries, http://novax.com/company/index.html

[2.20] Nooralahiyan A., Kirby H., and McKeown, D. Vehicle Classification by Acoustic Signature, Mathematical and Computer Modeling, vol. 27, no. 9-11, 1998, pp. 205-214.

[2.21] SmarTek SAS-1, http://www.smarteksys.com/sas-1_flyer.htm

[2.22] National Archives and Records Administration. U.S. GLOBAL POSITIONING SYSTEM POLICY. March 29, 1996.

[2.23] Liang Zou; Jian-Min Xu; Ling-Xiang Zhu; Arterial speed studies with taxi equipped with global positioning receivers as probe vehicle, Wireless Communications, Networking and Mobile Computing, 2005. Proceedings. 2005 International Conference on Volume 2, 23-26 Sept. 2005 Page(s):1343 - 1347

[2.24] GSM Localization, Wikipedia, http://en.wikipedia.org/wiki/GSM_localization

[2.25] Mark R. McCord, Yongliang Yang, Zhuojun Jiang, Benjamin Coifman, Prem K. Goel. Estimating AADT from Satellite Imagery and Air Photos: Empirical Results, Presented for 2003 Annual Meeting of Transportation Research Board, Washington D.C., January 2003.

[2.26] NIT Phase II Evaluation of Non-Intrusive Technologies for Traffic Detection, Final Report, Minnesota Department of Transportation, St. Paul, MN, September 2002.

[2.27] Nathan A. Weber, Verification of Radar Vehicle Detection Equipment, Report SD98-15-F, March 1999.

[2.28] Field Test of Monitoring of Urban Vehicle Operations Using Non-intrusive Technologies Final Report, Minnesota Department of Transportation, SRF Consulting Group, Inc., MN, May 1997.

[2.30] Yilin Zhao; Mobile phone location determination and its impact on intelligent transportation systems, Intelligent Transportation Systems, IEEE Transactions on Volume 1, Issue 1, March 2000 Page(s):55 - 64

[2.31] Wu Zhaohui; Liu Yanfei; Xu Wei; Mao Jian; Application of knowledge management in intelligent transportation systems, Intelligent Transportation Systems, 2003. Proceedings. 2003 IEEE Volume 2, 12-15 Oct. 2003 Page(s):1730 - 1734 vol.2

[2.32] Liu, Y.; Ozguner, U.; Acarman, T.; Performance evaluation of inter-vehicle communication in highway systems and in urban areas, Intelligent Transport Systems, IEE Proceedings Volume 153, Issue 1, March 2006 Page(s):63 - 75

[2.33] Kato, S., and Tsugawa, S.: 'Evaluation of information transmission over inter-vehicle communication with simulation studies'. Proc. IEEE 5th Int. Conf. Intelligent Transportation Systems, Singapore, September 2002, pp. 324–329

[3.1]   Nirupama Bulusu, Sanjay Jha; Wireless sensor networks, Publisher Boston, MA, Artech House, 2005.

[3.2]   MIT Technology Review 2003, http://www.techreview.com

[3.3]   Warneke, B., et al.; Smart Dust: Communicating with a Cubic-Millimeter Computer, IEEE Computer, Vol. 34, No. 1, January 2001, pp. 44-51

[3.4]   Cornelius T. Leondes; MEMS/NEMS: handbook, techniques and applications, Publisher New York: Springer, c2006.

[3.5]   Jason Lester Hill; System architecture for wireless sensor networks, Thesis (Ph.D. in Computer Science), University of California, Berkeley, Spring 2003.

[3.6]   A. Mainwaring, D. Culler, J. Polastre, R. Szewczyk, J. Anderson; Wireless sensor networks for habitat monitoring. In Proceedings of the 1st ACM International Workshop on Wireless Sensor Networks and Applications, pages 88-97, ACM Press, 2002

[3.7]   D. Niculescu and B. Nath; Ad hoc positioning system (APS), In Proceedings of GLOBE-COM'01 (IEEE), pages 2926-2931, San Antonio, Texas, November 2001.

[3.8]   Eyre, J.; Bier, J.; The evolution of DSP processors, Signal Processing Magazine, IEEE, Volume 17, Issue 2, March 2000 Page(s):43 - 51

[3.9]   Texas Instruments, MSP430X13x, MSP430x14x Mixed Signal Microcontroller User Guide. 2003: http://www-s.ti.com/sc/ds/msp430f149.pdf.

[3.10] Ergen, Sinem Coleri, Pravin Varaiya; PEDAMACS: Power Efficient and Delay Aware Medium Access Protocol for Sensor Networks, Mobile Computing, IEEE Transactions on Volume 5, Issue 7, July 2006 Page(s):920 – 930

[3.11] McLarnon, B.; VHF/UHF/Microwave Radio Propagation: A Primer for Digital Experimenters, http://www.tapr.org/tapr/html/ve3jf.dcc97/ve3jf.dcc97.html.

[3.12] Wilhelm, A., Surgenor, B.W., Pharoah, J.G.; Evaluation of a micro fuel cell as applied to a mobile robot, Mechatronics and Automation, 2005 IEEE International Conference, Volume 1, 29 July-1 Aug. 2005 Page(s):32 - 36

[3.13] Venkatasetty, H.V., Jeong, Y.U.; Recent advances in lithium-ion and lithium polymer batteries, Battery Conference on Applications and Advances, 2002. The Seventeenth Annual 15-18 Jan. 2002 Page(s):173 - 178

[3.14] Data sheet, HMC1051 / HMC1052 / HMC1053, 1,2 and 3-axes magnetic sensors, Honeywell International, http://www.honeywell.com/

[3.15] Data sheet, Crossbow Technology, MICA2DOT, http://www.xbow.com/

[3.16] Data sheet, TL-5135, TADIRAN Batteries, http://www.tadiranbat.com/

[3.17] SmartStud System, http://www.smartstud.com/

[3.18] Sensys Networks, Inc., http://www.sensysnetworks.com/

[3.19] Application note, AN218, Vehicle Detection Using AMR Sensors, Honeywell International, http://www.honeywell.com/

[3.20] TinyOS project, http://www.tinyos.net/

[3.21] Kalden, R., Meirick, I., Meyer, M.; Wireless Internet access based on GPRS, Personal Communications, IEEE, Volume 7, Issue 2, April 2000 Page(s):8 - 18

[3.22] Suwa, S., Atarashi, H., Sawahashi, M.; Performance comparison between MC/DS-CDMA and MC-CDMA for reverse link broadband packet wireless access, Vehicular Technology Conference, 2002. Proceedings. VTC 2002-Fall. 2002 IEEE 56th, Volume 4, 24-28 Sept. 2002 Page(s):2076 - 2080 vol.4

[3.23] Atmel Corporation, http://www.atmel.com/

[3.24] Chipcon, CC1000 Single Chip Very Low Power RF Transceiver. 2002, http://www.chipcon.com/files/CC1000_Data_Sheet_2_1.pdf.

[3.25] Handbook of wireless networks and mobile computing, edited by Ivan Stojmenović. Published by New York : Wiley, 2002.

[3.26] W. Ye, J. Heidemann, and D. Estrin, An Energy-Efficient MAC Protocol for Wireless Sensor Networks,IEEE INFOCOM 2002, June 2002

[3.27] C. S. Raghavendra and S. Singh, PAMAS - Power Aware Multi-Access Protocol with Signalling for Ad Hoc Networks, Computer Communications Review, July 1998

[3.28] R. Ramaswami and K. K. Parhi, Disributed Scheduling of Broadcasts in a Radio Network, Proceedings of the Eighth Annual Joint Conference of the IEEE Computer and Communications Societies. Technology: Emerging or Converging, IEEE , 1989 pp. 497 - 504 vol.2, INFOCOM 1989.

[3.29] Simulation with TOSSIM, http://www.tinyos.net/tinyos-2.x/doc/html/tutorial/lesson11.html

[3.30] J. Polastre, R. Szewczyk, C. Sharp and D. Culler, The Mote Revolution: Low Power Wireless Sensor Network Devices, Hot Chips 16: A symposium on High Performance Chips, August 2004.

[3.31] S. Coleri and P. Varaiya; "Optimal Placement of Relay Nodes in Sensor Networks," Technical Report, Department of EECS, University of California, Berkeley

[4.1] Ding, Jiagen; Cheung, Sing Yiu.; Tan, Chin Woo.; Varaiya, Pravin; Signal processing of sensor node data for vehicle detection, Intelligent Transportation Systems,

2004. Proceedings. The 7th International IEEE Conference on 3-6 Oct. 2004 Page(s):70 - 75

[4.2]   Data sheet, Crossbow Technology, MICA2 (MPR410),

http://www.xbow.com/Products/Product_pdf_files/Wireless_pdf/MICA2_Datasheet.pdf

[4.3]   Data sheet, Crossbow Technology, Multi sensor board (MTS310),

http://www.xbow.com/Products/Product_pdf_files/Wireless_pdf/MTS_MDA_Datasheet.pdf

[4.4]   Data sheet, Crossbow Technology, Multi sensor board (MTS510),

http://www.xbow.com/Products/Product_pdf_files/Wireless_pdf/MTS510CA_Datasheet.pdf

[4.5]   Data sheet, Panasonic WM-62A microphone,

http://www.panasonic.com/industrial/components/pdf/em05_wm62_a_c_cc_k_b_dne.pdf

[4.6]   Pederson, M.; Olthuis, W.; Bergveld, P.; High-performance condenser microphone with fully integrated CMOS amplifier and DC-DC voltage converter, Microelectromechanical Systems, Journal of Volume 7, Issue 4, Dec. 1998 Page(s):387 - 394

[4.7]   Sorensen, H.V.; Burrus, C.S.; Efficient computation of the short-time fast Fourier transform, Acoustics, Speech, and Signal Processing, 1988. ICASSP-88., 1988 International Conference on 11-14 April 1988 Page(s):1894 - 1897 vol.3

[4.8]   SFFT documentation for MATLAB , http://www.mathworks.com/ access/helpdesk/help/toolbox/dspblks/shorttimefft.shtml

[4.9]   Robert J. Schilling and Sandra L. Harris; Fundamentals of digital signal processing using MATLAB, Southbank, Vic. , United States, c2005.

[4.10] Sing Yiu Cheung, Sinem Coleri, Baris Dundar, Sumitra Ganesh, Chin-Woo Tan, Pravin Varaiya; Traffic Measurement and Vehicle Classification with a Single Magnetic Sensor, PATH Report UCB-ITS-PWP-2004-07, 2004

[4.11] B. Coifman, S. Dhoorjaty and Z.-H. Lee. Estimating median velocity instead of mean velocity at single loop detectors. Transportation Research, Part C, vol 11C, nos. 3-4, pp. 211-222, June-August 2003.

[4.12] Jiagen(Jason) Ding, Sing-Yiu Cheung, Chin-woo Tan and Pravin Varaiya; Vehicle Detection by Sensor Network Nodes, PATH Report UCB-ITS-PRR-2004-39, 2004

[4.13] Peter T. Martin, Yuqi Feng, Xiaodong Wang; Detector Technology Evaluation, Department of Civil and Environmental Engineering, University of Utah Traffic Lab, report MPC03-154, Nov 2003

[4.14] C. Oh, S.G. Ritchie and S.-T. Jeng. Vehicle re-identification using heterogeneous detection systems. 83rd TRB Annual Meeting, January 2004, Washington, D.C.

[4.15] S. Oh, S.G. Ritchie and C. Oh. Real time traffic measurement from single loop inductive signatures. 81st TRB Annual Meeting, January 2002, Washington, D.C.

[5.1]   Wei, C. et al. Vehicle Classification Using Advanced Technologies. Transportation Research Record 1551. November. 1996. pp. 45-50.

[5.2]   S. Gupte, et al., Detection and Classification of Vehicles, IEEE Trans. on ITS, Vol. 3, No. 1, 2002, pp.37-47.

[5.3]   D. Koller, Moving object recognition and classification based on recursive shape parameter estimation, in Proc. 12th Israel Conf. Artificial Intelligence, Computer Vision, Dec. 27–28, 1993.

214

[5.4]   G. D. Sullivan, "Model-based vision for traffic scenes using the groundplane constraint," Phil. Trans. Roy. Soc. (B), vol. 337, pp. 361–370, 1992.

[5.5]   G. D. Sullivan, K. D. Baker, A. D.Worrall, C. I. Attwood, and P. M. Remagnino, Model-based vehicle detection and classification using orthographic approximations, Image Vis. Comput., vol. 15, no. 8, pp. 649–654, Aug. 1997.

[5.6]   Yuan, X. et al. Computer Vision System for Automatic Vehicle Classification. ASCE Journal of Transportation Engineering. Vol. 120. No. 6. November/December. 1994. Pp. 861-876.

[5.7]   Carlos Sun. An Investigation in the Use of Inductive Loop Signatures for Vehicle Classification, California PATH Research Report, UCB-ITS-PRR-2000-4.

[5.8]   Pursula, M. and Pikkarainen, P. A Neural Network Approach to Vehicle Classification with Double Induction Loops. Proceedings of the 17th ARRB Conference. Part 4. 1994. pp. 29-44.

[5.9]   M. Bishop, Neural Networks for Pattern Recognition. New York: Oxford, 1995.

[5.10] T. Kohonen, J. Kangas, J. Laaksonen, K. Torkkola, LVQPAK: The Learning Vector Quantization Program Package, Technical Report from the LVQ Programming Team of the Helsinki University of Technology, Laboratory of Computer and Information Science, Rakentajanaukio 2C, SF-02150 ESPOOF, inland, 31 January 1992

[5.11] USGS National Geomagnetism Program, http://geomag.usgs.gov/

[5.12] J. Kwon, A. Skabardonis and P. Varaiya. Estimation of truck traffic volume from single loop detector using lane-to-lane speed correlation. 82nd TRB Annual Meeting, January 2003, Washington, D.C.

[5.13]   X. Zhang, Y. Wang and N.L. Nihan. Monitoring a freeway network in real-time using single-loop detectors: System design and implementation. 83rd TRB Annual Meeting, January 2004, Washington, D.C.

[5.14]   Chen, C.S. and Huo, K.S. Karhunen-Loeve Method for Data Compression and Speech Synthesis. IEEE Proceedings-I, Vol. 138, No. 5, October. Pp. 377-380.1991.

[5.15]   Abbas, H.M. and Fahmy, M.M. Neural Model for Karhunen-Loeve Transform with Application to Adaptive Image Compression. IEE Proceedings-I, Vol. 140, No. 2, April. Pp. 135-143. 1993.

[5.16]   Strang G (1998). "Introduction to Linear Algebra". Section 6.7. 3rd ed., Wellesley-Cambridge Press. ISBN 0961408855

[5.17]   T. M. Cover and P. E. Hart (1967), Nearest Neighbor Pattern Classification, in Trans. IEEE Inform. Theory, IT-13, pp21-27.

[5.18]   T. M. Cover (1968), Rates of Convergence for Nearest Neighbor Procedures, in Proc. Hawaii Int. Conf. on System Science, pp413-415.

[5.19]   Yu Jiangsheng,Method of k-Nearest Neighbors, Institute of Computational Linguistics, Peking University, China, 100871, September 3, 2002

[5.20]   Bin Zhang; Srihari, S.N.; Fast k-nearest neighbor classification using cluster-based trees,Pattern Analysis and Machine Intelligence, IEEE Transactions on Volume 26, Issue 4, April 2004 Page(s):525 - 528

[5.21]   B.E. Boser et al. A Training Algorithm for Optimal Margin Classifiers. Proceedings of the Fifth Annual Workshop on Computational Learning Theory 5 144-152, Pittsburgh, 1992

[5.22] L. Bottou et al. Comparison of classifier methods: a case study in handwritten digit recognition. Proceedings of the 12th IAPR International Conference on Pattern Recognition, vol. 2, pp. 77-82.

[5.23] J. Platt, Sequential Minimal Optimization: A Fast Algorithm for Training Support Vector Machines, Microsoft Research Technical Report MSR-TR-98-14, (1998).

[5.24] C. Cortes and V. Vapnik, Support-Vector Networks, Machine Learning, 20(3):273-297, September 1995 http://citeseer.ist.psu.edu/cortes95supportvector.html

[5.25] B. E. Boser, I. M. Guyon, and V. N. Vapnik. A training algorithm for optimal margin classifiers. In D. Haussler, editor, 5th Annual ACM Workshop on COLT, pages 144-152, Pittsburgh, PA, 1992. ACM Press

[5.26] Bernhard Scholkopf, Christopher J.C. Burges, and Alexander J. Smola (editors). "Advances in Kernel Methods: Support Vector Learning". MIT Press, Cambridge, MA, 1999

[5.27] Hsu, C. W. et al. A comparison of methods for multi-class support vector machines, IEEE Transactions on Neural Networks. Vol. 13. pp. 415-425.

[5.28] Janusz Gajda, Ryszard Sroka, Marek Stencel, Andrzej Wajda, Tadeusz Zeglen, A Vehicle Classification Based on Inductive Loop Detectors, IEEE Instrumentation and Measurement Technology Conference, Budapest, Hungary, May 21-23,2001.

[6.1]   C. Sun, S. Ritchie, K. Tsai, and R. Jayakrishnan, "Use of vehicle signature analysis and lexicographic optimization for vehicle reidentification on freeways," Transport. Res., pt. C, vol. 7, pp. 167–185, 1999.

[6.2]   Cheol Oh; Ritchie, S.G.; Jun-Seok Oh; Jayakrishnan, R.; Real-time origin-destination (OD) estimation via anonymous vehicle tracking Intelligent Transportation Systems, 2002. Proceedings. The IEEE 5th International Conference on 2002 Page(s):582 - 586

[6.3]   K. F. Petty et al., "Accurate estimation of travel times from single-loop detectors," in Proc. 76th Annu. Transportation Research Board Meeting, Washington, DC, Jan. 1997.

[6.4]   Huairui Guo, Jionghua Jin; "Travel Time Estimation Using Correlation Analysis of Single Loop Detector Data" Transport. Res. B, 2006

[6.5]   B. Coifman, "Vehicle re-identification and travel time measurement in real-time on freeways using the existing loop detector infrastructure," presented at the 77th Annu. Transportation Research Board Meeting, Washington, DC, 1998.

[6.6]   Turner, S., W. Eisele, R. Benz, and D. Holdener. Travel Time Data Collection Handbook. FHWA-PL-98-035. Federal Highway Administration and Texas Transportation Institute. 1998.

[6.7]   Harry H. Cheng, Ben Shaw, Joe Palen, Zhaoqing Wang, Bo Chen, A Field-Deployable Real-Time Laser-Based Non-Intrusive Detection System for Measurement of True Travel Time on the Highway University of California, Davis, California PATH Research Report, UCB-ITS-PRR-2002-23

[6.8]   Urazghildiiev, I.R.; Ragnarsson, R.; Wallin, K.; Rydberg, A.; Ridderstrom, P.; Ojefors, E.; A vehicle classification system based on microwave radar measurement of height profiles, RADAR 2002 15-17 Oct. 2002 Page(s):409 - 413

[6.9]   Christiansen, I, Hauer, L. "Probing for Travel Time: Norway Applies AVI and

WIM Technologies for Section Probe Data", Traffic Technology International, Aug/Sep

1996, UK & International Press, Surrey, UK, 1996, pp 41-44.

[6.10]   MacCarley, C.A. Video-Based Vehicle Signature Analysis and Tracking System

Phase 2: Algorithm Development and Preliminary Testing. MOU 350 Final Report to

California PATH, California Polytechnique State University, San Luis Obispo. 2001.

[6.11]   Chung-Lin Huang; Wen-Chieh Liao; A vision-based vehicle identification system,

Pattern Recognition, 2004. ICPR 2004. Proceedings of the 17th International Conference

on Volume 4, 23-26 Aug. 2004 Page(s):364 - 367 Vol.4

[6.12]   Ritchie, S.G., and C. Sun. Section Related Measures of Traffic System

Performance: Final Report. California PATH Research Report UCB-ITS-PRR-98-33.

1998.

[6.13]   Sun, C., S.G. Ritchie, and K. Tsai. Algorithm Development for Derivation of

Section-Related Measures of Traffic System Performance using Inductive Loop Detectors.

In Transportation Research Record 1643, TRB, National Research Council, Washington,

D.C., 1998, pp. 171-180.

[6.14]   Oh, C., S.G. Ritchie, S. Park. Development A Real-Time Probabilistic Vehicle

Reidentification Algorithm for Signalized Intersections. Working paper, Institute of

Transportation Studies, University of California, Irvine, 2002.

[6.15]   Stephen G. Ritchie, Seri Park, Cheol Oh, Shin-Ting (Cindy) Jeng, Andre Tok; Field

Investigation of Advanced Vehicle Reidentification Techniques and Detector Technologies

– Phase 2, California PATH Research Report, UCB-ITS-PRR-2005-8

[6.16] Yanying Li; McDonald, M.; Link travel time estimation using single GPS equipped probe vehicle, Intelligent Transportation Systems, 2002. Proceedings. The IEEE 5th International Conference on 2002 Page(s):932 - 937

[6.17] Xuesong Zhou; Mahmassani, H.S.; Dynamic origin-destination demand estimation using automatic vehicle identification data, Intelligent Transportation Systems, IEEE Transactions on Volume 7, Issue 1, March 2006 Page(s):105 - 114

[6.18] Ygnace, J.-L.; Drane, C.; Cellular telecommunication and transportation convergence: a case study of a research conducted in California and in France on cellular positioning techniques and transportation issues Intelligent Transportation Systems, 2001. Proceedings. 2001 IEEE 25-29 Aug. 2001 Page(s):16 - 22

[6.19] V. D. Zijpp, "Dynamic origin-destination matrix estimation on motorway networks," Ph. D. dissertation, Dept. Transport. Planning Traffic Eng., Delft Univ. Technology, GA Delft, Netherlands, 1996.

[6.20] C. Oh, S.G. Ritchie, J. Oh and R. Jayakrishnan, "Real-time origin-destination (OD) estimation via anonymous vehicle tracking," in Proc. IEEE Int. Conf. Intell. Transport. Syst., Singapore, 2002, pp. 582 – 586.

[7.1] S. C. Wong and H. Lo, "Advanced algorithms in traffic signal control: Editorial," J. Intell. Transp. Syst., vol. 8, no. 2, pp. 61–62, Apr.–Jun. 2004.

[7.2] York, I.; Al-Katib, M.; Methods of traffic signal control and signal timings at high speed sites, Road Transport Information and Control, 2000. Tenth International Conference on (Conf. Publ. No. 472) 4-6 April 2000 Page(s):167 - 171

[7.3]   Bonneson, J.A., D.R. Middleton, K.H. Zimmerman, H.A. Charara, and M.M.

Abbas. New Detection System for Rural Signalized Intersections. Report No. TX-00/4022-

S. Texas Department of Transportation, Austin, Texas, August 2000.

[7.4]   Nadeem, C.A., M.M. Abbas, H.A. Charara, and R.T. Parker. An Intelligent Control

System for Detecting and Progressing Platoons at Isolated Traffic Signals. Report No. TX-

03/4304-S. Texas Department of Transportation, Austin, Texas, September 2003.

[7.5]   Messer, C.J., S.R. Sunkari, H.A. Charara, and R.T. Parker. Advance Warning for

End-of-Green to Reduce Red-Light-Running. Report No. TX-03/4260-S, Texas

Department of Transportation, Austin, Texas, September 2003.

[7.6]   Srinivasa Sunkari, Ricky Parker, Hassan Charara, Trishul Palekar; Evaluation of

cost-effective technologies for advanced detection, Texas Transportation Institute, Report

FHWA/TX-06/0-5002-1

[7.7]   VDS240 for Advance Detection, Application notes, Sensys Networks Inc.,

http://www.sensysnetworks.com/vehicle_detection_applications.html

[7.8]   Twin Cities Ramp Meter Evaluation, Cambridge Systematics, Inc. February 2001.

[7.9]   Cambridge Systematics, Inc. and Texas Transportation Institute, "Traffic

congestion and reliability: Linking solutions to problems," Federal Highway

Administration, Tech. Rep., July 19, 2004.

http://www.ops.fhwa.dot.gov/congestion_report/

[7.10]   Xiaotian Sun; Horowitz, R.; A localized switching ramp-metering controller with a

queue length regulator for congested freeways, American Control Conference, 2005.

Proceedings of the 2005 8-10 June 2005 Page(s):2141 - 2146 vol. 3

[7.11]   E. Smaragdis and M. Papageorgiou, "Series of new local ramp metering strategies," Transportation Research Record, no. 1856, pp. 74–86, 2004.

[7.12]   R. L. Gordon, "Algorithm for controlling spillback from ramp meters," Transportation Research Record, no. 1554, pp. 162–171, 1996.

[7.13]   Gabriel C.Gomes; Optimization and Microsimulation of On-ramp Metering for Congested Freeways, California PATH research report, UCB-ITS-PRR-2004-44

[7.14]   Sakai, A.; Mizuno, K.; Sugimoto, T.; Okuda, T.;Parking guidance and information systems, Vehicle Navigation and Information Systems Conference, 1995. Proceedings. In conjunction with the Pacific Rim TransTech Conference. 6th International VNIS. 'A Ride into the Future',30 July-2 Aug. 1995 Page(s):478 - 485

[7.15]   Griffith, E. Pointing the way. ITS International, March/April 2000. p. 72.

[7.16]   Susan Shaheen, Caroline Rodier, Amanda M. Eaken; Smart Parking Management Pilot Project: A Bay Area Rapid Transit (BART) District Parking Demonstration, California PATH Research Report, UCB-ITS-PRR-2005-5, 2005

[7.17]   Merriman. D. How many parking spaces does it take to create one additional transit passenger? Regional Science and Urban Economics, Vol. 28, pp. 565-584. 1998

[7.18]   Burdette, D. An evaluation of advanced parking information systems at airports. Preprint. Transportation Research Board Annual Meeting, Washington D.C. 2001

[7.19]   PARC240 Accurate Lane Count System, Sensys Networks Inc., http://www.sensysnetworks.com/vehicle_detection_applications.html

[7.20]   Highway Statistics 2000, FHWA Report, http://www.fhwa.dot.gov/ohim/hs00/index.htm.

[7.21] Intelligent Transportation Systems in Work Zones - A Cross-Cutting Study: Integrated Work Zone Systems for Improving Travel Conditions and Safety, FHWA Report FHWA-OP-02-025, 2002

[7.22] Kajija Zasuhiko, Ishimoto Keishi, Matsuzawa Masaru, "ITS Technology for Winter Road Management" , PIARC International Winter Road Congress, Lulea, Sweden March 1998

[7.23] Ask, Bernie, "Implementing an Integrated Road Weather Information System in Minesota", PIARC International Winter Road Congress, Osaka, Japan May 2002

[7.24] MICA weather board, MTS420/400, data sheet, Crossbow Technology, http://www.xbow.com/Products/Product_pdf_files/Wireless_pdf/MTS400-420_Datasheet.pdf

[7.25] Clarus - America's 21st Century Surface Transportation Weather Observing and Forecasting System, Federal Highway Administration, U.S. Department of Transportation, http://www.clarusinitiative.org/

[7.26] Traffic Congestion and Reliability: Linking Solutions to Problems. Cambridge Systematics, Inc. and Texas Transportation Institute. July 2004

[7.27] Vehicle Infrasttucture Integration (VII) Fact Sheet, Intelligent Transportation Systems, U.S. Department of Transportation, http://www.its.dot.gov/vii/index.htm

[7.28] Cseh, C.;Architecture of the dedicated short-range communications (DSRC) protocol, Vehicular Technology Conference, 1998. VTC 98. 48th IEEE Volume 3, 18-21 May 1998 Page(s):2095 - 2099 vol.3

[7.29] Ching-Yao Chan; California Intersection Decision Support: A Systems Approach to Achieve Nationally Interoperable Solutions, California PATH Research Report, UCB-ITS-PRR-2005-11

[8.1] S. Roundy, Energy Scavenging for Wireless Sensor Nodes with a Focus on Vibration to Electricity Conversion. Ph. D. Dissertation, Dept. of EECS, UC Berkeley, May 2003.

[8.2] K. Lahiri, A. Raghunathan, and S. Dey, "Battery-driven system design: a new frontier in low power design", in Proc. IEEE International Conference on VLSI Design, pp. 261–267, 2002.

[8.3] Vijay Raghunathan; Kansal, A.; Hsu, J.; Friedman, J.; Mani Srivastava; Design considerations for solar energy harvesting wireless embedded systems, Information Processing in Sensor Networks, 2005. IPSN 2005. Fourth International Symposium on 15 April 2005 Page(s):457 - 462

www.ingramcontent.com/pod-product-compliance
Lightning Source LLC
LaVergne TN
LVHW022306060326
832902LV00020B/3310